was proved under restrictions that are not valid for the game of this paper, at first we reconsider this result under less restrictive conditions and give an easier proof of the extended separation principle than it was given in [6]. Then we study the saddle point in the linear quadratic game of two players, each having individual observations under the dependent noises.

Two cases are considered. If the observations of one of the players include the observations of the other player, then we prove the representation for the optimal strategy of the player with worse observations. In the case when the players have the same observations, we prove the representation for the saddle point. We also discuss the existence of a saddle point in two games over the heat and wave equations.

2. Notation

In this paper it is always assumed that $(\Omega, \mathcal{F}, \mathbf{P})$ is a complete probability space, $X, Y, Z, Z_1, Z_2, U, U_1, U_2$ are separable Hilbert spaces, $T > 0$, $\mathbf{T} = [0, T]$ is a finite time interval and $\Delta_T = \{(t, s) : 0 \le s \le t \le T\}$. The space of bounded linear operators from X to Z is denoted by $\mathcal{L}(X, Z)$. The brief notation $\mathcal{L}(X) = \mathcal{L}(X, X)$ is used as well. A^* denotes the adjoint of the operator A. If $A \in \mathcal{L}(X)$ is self-adjoint and $\langle h, Ah\rangle \ge 0$ (respectively, $\langle h, Ah\rangle \ge c\|h\|^2$ with $c = \text{const.} > 0$) for all $h \in X$, then we write $A \ge 0$ (respectively, $A > 0$), where $\langle \cdot, \cdot\rangle$ is a scalar product and $\|\cdot\|$ is a norm. The identity operators will be denoted by I and the zero operators by 0.

For the operators $A \in \mathcal{L}(X \times Z)$, $B \in \mathcal{L}(U, X \times Z)$, $C \in \mathcal{L}(X \times Z, U)$ and for the vector $h \in X \times Z$, we use the decompositions

$$A = \begin{bmatrix} A_1 & A_2 \\ A_3 & A_4 \end{bmatrix}, \ B = \begin{bmatrix} B_1 \\ B_2 \end{bmatrix}, \ C = \begin{bmatrix} C_1 & C_2 \end{bmatrix}, \ h = \begin{bmatrix} h_1 \\ h_2 \end{bmatrix},$$

where $A_1 \in \mathcal{L}(X)$, $A_2 \in \mathcal{L}(Z, X)$, $A_3 \in \mathcal{L}(X, Z)$, $A_4 \in \mathcal{L}(Z)$, $B_1 \in \mathcal{L}(U, X)$, $B_2 \in \mathcal{L}(U, Z)$, $C_1 \in \mathcal{L}(X, U)$, $C_2 \in \mathcal{L}(Z, U)$, $h_1 \in X$ and $h_2 \in Z$.

We refer to Hille and Phillips [7] for the concepts of continuity and measurability of vector- and operator-valued functions. Let (S, Σ, ν) be a measure space. By $L_2(S, \Sigma, \nu, X)$ ($L_\infty(S, \Sigma, \nu, X)$) we denote the space of equivalence classes of strongly Σ-measurable and square ν-integrable (essentially bounded) vector-valued functions from S to X. Respectively, we denote by $B_2(S, \Sigma, \nu, \mathcal{L}(X, Z))$ ($B_\infty(S, \Sigma, \nu, \mathcal{L}(X, Z))$) the space of equivalence classes of strongly Σ-measurable and square ν-integrable (essentially bounded) operator-valued functions from S to $\mathcal{L}(X, Z)$. If $(S, \Sigma, \nu) = (\Omega, \mathcal{F}, \mathbf{P})$ or $(S, \Sigma, \nu) = (G, \mathcal{L}_G, \ell)$, where G is a Lebesgue measurable subset of a finite dimensional Euclidean space, $\mathcal{L}_G$ is the σ-algebra of Lebesgue measurable subsets of G and ℓ is the Lebesgue measure on G, then the σ-algebra Σ and the measure ν will be dropped in the above mentioned notation. Also, we drop X in the notation $L_2(S, \Sigma, \nu, X)$ if X is the real line that is denoted by $\mathbb{R}$. By $C(\mathbf{T}, X)$ we denote the space of strongly continuous functions from $\mathbf{T}$

to X and by $B(\mathbf{T},\mathcal{L}(X,Z))$ the class of strongly Borel measurable and bounded operator-valued functions from $\mathbf{T}$ to $\mathcal{L}(X,Z)$.

A function $\mathcal{U}:\Delta_T\to\mathcal{L}(X)$ is called a *mild evolution operator* (see Curtain and Ichikawa [5]) if

(a) $\mathcal{U}_{t,t}=I,\ 0\le t\le T$;
(b) $\mathcal{U}_{t,s}=\mathcal{U}_{t,r}\mathcal{U}_{r,s},\ 0\le s\le r\le t\le T$;
(c) $\mathcal{U}_{t,\cdot}:[0,t]\to\mathcal{L}(X)$ and $\mathcal{U}_{\cdot,s}:[s,T]\to\mathcal{L}(X)$ are weakly continuous for all $0<t\le T$ and for all $0\le s<T$;
(d) $\sup_{\Delta_T}\|\mathcal{U}_{t,s}\|<\infty$.

The class of all mild evolution operators from Δ_T to $\mathcal{L}(X)$ will be denoted by $\mathcal{E}(\Delta_T,\mathcal{L}(X))$. For $\mathcal{U}\in\mathcal{E}(\Delta_T,\mathcal{L}(X))$ and $N\in B_\infty(\mathbf{T},\mathcal{L}(X))$, the equation

$$\mathcal{Y}_{t,s}=\mathcal{U}_{t,s}+\int_s^t\mathcal{Y}_{t,r}N_r\mathcal{U}_{r,s}\,dr,\ \ 0\le s\le t\le T,$$

where the integral is in the strong Bochner sense, has a unique solution in $\mathcal{E}(\Delta_T,\mathcal{L}(X))$. This solution is called a *bounded perturbation* of $\mathcal{U}$ by N and it will be denoted by $\mathcal{Y}=\mathcal{P}_N(\mathcal{U})$.

An X-valued random variable is a strongly $\mathcal{F}$-measurable function from Ω to X and the collection of them will be denoted by $m(\Omega,X)$. The space of X-valued square integrable martingales on $\mathbf{T}$ and its subspace consisting of martingales with continuous paths will be denoted by $M_2(\mathbf{T},X)$ and $M_2^c(\mathbf{T},X)$, respectively. By $\sigma(\xi_t;a\le t\le b)$ we denote the σ-algebra generated by the random variables ξ_t, $a\le t\le b$. We will use the following notation as well: for $h\in L_2(\mathbf{T}\times\Omega;Z)$ and for $0<t<T$, we denote by h^t and h^{t+} the $L_2(0,t;Z)$- and $L_2(t,T;Z)$-valued random variables obtained by restriction of h to $[0,t]\times\Omega$ and to $(t,T]\times\Omega$, respectively.

All integrals of vector-valued functions are considered in the Bochner sense. For expectation and for conditional expectation, the notation $\mathbf{E}$ and $\mathbf{E}(\cdot|\cdot)$ is used, respectively. The integrals of operator-valued functions (except stochastic integrals) are in the strong Bochner sense. The stochastic integral of the function $\Phi\in B(\mathbf{T},\mathcal{L}(X,Z))$ with respect to $m\in M_2(\mathbf{T},X)$ is well defined. Moreover, if $\Psi\in B(\mathbf{T}\times\mathbf{T},\mathcal{L}(X,Z))$, then the equality

$$\int_0^T\int_0^T\Phi_{t,s}\,dm_t\,ds=\int_0^T\int_0^T\Phi_{t,s}\,ds\,dm_t$$

holds with probability 1 (the stochastic analogue of Fubini's theorem) which will be used in the sequel.

3. Control Problem

3.1. Setting of the Problem

Consider the linear state-observation system

$$x_t^u = \mathcal{U}_{t,0} l + \int_0^t \mathcal{U}_{t,s}(B_s u_s + b_s)\, ds + \int_0^t \mathcal{U}_{t,s}\, dm_s, \ 0 \le t \le T, \tag{3.1}$$

$$z_t^u = \int_0^t (C_s x_s^u + c_s)\, ds + n_t, \ 0 \le t \le T, \tag{3.2}$$

and the functional

$$J(u) = \mathbf{E}\Bigg(\langle x_T^u, Q_T x_T^u \rangle + \int_0^T \left\langle \begin{bmatrix} x_t^u \\ u_t \end{bmatrix}, \begin{bmatrix} F_t & L_t^* \\ L_t & G_t \end{bmatrix} \begin{bmatrix} x_t^u \\ u_t \end{bmatrix} \right\rangle dt$$
$$+ 2\langle q, x_T^u \rangle + 2 \int_0^T \left\langle \begin{bmatrix} f_t \\ g_t \end{bmatrix}, \begin{bmatrix} x_t^u \\ u_t \end{bmatrix} \right\rangle dt \Bigg) \tag{3.3}$$

where u is a control taken from the set of admissible controls U_{ad}, which will be defined below, and x^u and z^u are the state and observation processes corresponding to the control u. The following conditions are supposed to hold:

($\mathbf{C_1}$) $\mathcal{U} \in \mathcal{E}(\Delta_T, \mathcal{L}(X))$, $B \in B_\infty(\mathbf{T}, \mathcal{L}(U, X))$, $C \in B_\infty(\mathbf{T}, \mathcal{L}(X, Z))$;

($\mathbf{C_2}$) $l \in L_2(\Omega, X)$, $b \in L_2(\mathbf{T} \times \Omega, X)$, $c \in L_2(\mathbf{T} \times \Omega, Z)$, $m \in M_2(\mathbf{T}, X)$, $n \in M_2(\mathbf{T}, Z)$;

($\mathbf{C_3}$) $Q_T \in \mathcal{L}(X)$, $Q_T^* = Q_T$, $F \in B_\infty(\mathbf{T}, \mathcal{L}(X))$, $G, G^{-1} \in B_\infty(\mathbf{T}, \mathcal{L}(U))$, $L \in B_\infty(\mathbf{T}, \mathcal{L}(X, U))$, $F_t^* = F_t$ and $G_t^* = G_t$ for a.e. $t \in \mathbf{T}$;

($\mathbf{C_4}$) $q \in L_2(\Omega, X)$, $f \in L_2(\mathbf{T} \times \Omega, X)$, $g \in L_2(\mathbf{T} \times \Omega, U)$.

For given $u \in L_2(\mathbf{T}\times\Omega, U)$, the state system (3.1) defines the random process x^u in $L_\infty(\mathbf{T}, L_2(\Omega, X))$. Hence, the random process z^u, defined by (3.2), belongs to $m(\Omega, L_2(\mathbf{T}, Z))$. According to our convention from Section 2, for $0 < t \le T$, the restriction of z^u to $[0, t] \times \Omega$, considered as an $L_2(0, t; Z)$-valued random variable, is denoted by $z^{u,t}$. Consider

$$U_t^u = L_2\big(\Omega, \sigma(z^{u,t}), \mathbf{P}, U\big).$$

We denote $z^{0,t} = z^{u,t}$ and $U_t^0 = U_t^u$ if $u = 0$. Let

$$U^0 = \int_0^T U_t^0\, dt = \big\{u \in L_2(\mathbf{T} \times \Omega, U) : u_t \in U_t^0 \text{ for a.e. } t \in \mathbf{T}\big\},$$
$$U^1 = \{u \in L_2(\mathbf{T} \times \Omega, U) : u_t \in U_t^u \text{ for a.e. } t \in \mathbf{T}\},$$

where the integral is a Hilbertian sum of subspaces. For any fixed control $u \in L_2(\mathbf{T}\times\Omega, U)$, $\{U_t^u : 0 < t \le T\}$ is a nondecreasing family of subspaces of $L_2(\Omega, U)$. Additionally, U^0 is a subspace of $L_2(\mathbf{T}\times\Omega, U)$. But, we can not say the same about U^1.

Since U^1 consists of functions $u \in L_2(\mathbf{T} \times \Omega, U)$ adapted with respect to the partial observations, it could be chosen as a set of admissible controls. But, the complicated structure of U^1 makes it inconvenient in studying optimal control problems. On the other hand, U^0 has an excellent structure, but the choice of U^0 as a set of admissible controls means disregarding the partial observations. Hence, following Bensoussan and Viot [4], the set of admissible controls will be defined by

$$U_{\mathrm{ad}} = U^0 \cap U^1. \tag{3.4}$$

Note that $U_{\mathrm{ad}} \neq \varnothing$ since at least it contains the nonrandom controls. The properties of U_{ad}, presented below, show that U_{ad} is a good trade-off U^1 against U^0.

Proposition 3.1. *Under the above conditions, the following statements hold:*

(a) *If $u \in U_{\mathrm{ad}}$, then $U_t^u = U_t^0$, $0 < t \le T$.*
(b) *If $\varepsilon > 0$ and if $u \in \int_0^T U_{t-\varepsilon}^u \, dt$, where $U_{t-\varepsilon}^u = U$ for $0 \le t \le \varepsilon$, then $u \in U_{\mathrm{ad}}$.*
(c) *U_{ad} is dense in U^0.*
(d) *$\inf_{U_{\mathrm{ad}}} J(u) = \inf_{U^0} J(u)$.*

Proof. In Bensoussan and Viot [4] this proposition is proved under the condition $z^u \in m(\Omega, C(\mathbf{T}, Z))$. This proof with slight changes remains valid in the case $z^u \in m(\Omega, L_2(\mathbf{T}, Z))$. □

Remark 3.2. In Bensoussan and Viot [4] it is shown that the condition $u \in U^1$ is not sufficient for $U_t^u = U_t^0$, $0 < t \le T$. Since the latter equalities are essential in the sequel, the set U^1 is inconvenient as a set of admissible controls. On the other hand, by Proposition 3.1(b), the set of admissible controls U_{ad} contains all functions in $L_2(\mathbf{T} \times \Omega, U)$ which are adapted with respect to partial observations up to an arbitrarily small delay of current time. Since in practice, for real reasons, synthesizing of controls at current time takes away some time for processing observation data, the use of $U^1 \cap U^0$ as a set of admissible controls rather than U^1 is quite reasonable.

Remark 3.3. If in the condition $(\mathbf{C_2})$ we take $n \in M_2^c(\mathbf{T}, Z)$ instead of $n \in M_2(\mathbf{T}, Z)$, then $z^u \in m(\Omega, C(\mathbf{T}, Z)) \subset m(\Omega, L_2(\mathbf{T}, Z))$. Let $z^{c,u,t}$ and $z^{u,t}$ be random variables with values in $C(0,t;Z)$ and $L_2(0,t;Z)$, respectively, obtained by restriction of z^u from $\mathbf{T} \times \Omega$ to $[0,t] \times \Omega$. Similar to the definitions of U_t^u, U^0 and U^1, one can define

$$U_t^{c,u} = L_2\left(\Omega, \sigma\left(z^{c,u,t}\right), \mathbf{P}, U\right), \; U^{c,0} = \int_0^T U_t^{c,0} \, dt,$$
$$U^{c,1} = \{u \in L_2(\mathbf{T} \times \Omega, U) : u_t \in U_t^{c,u} \text{ for a.e. } t \in \mathbf{T}\}.$$

In Bensoussan and Viot [4] (as well as in Curtain and Ichikawa [5] and in Bashirov [6]) the set $U_{\mathrm{ad}}^c = U^{c,0} \cap U^{c,1}$ is taken as a set of admissible controls. We clearly have $U_{\mathrm{ad}}^c = U_{\mathrm{ad}}$, i.e., the sets U_{ad}^c and U_{ad} are equal. But, U_{ad}^c can be defined only for observations with continuous paths, while the definition of U_{ad} does not use the continuity of observation paths. Therefore, U_{ad} is an extension of U_{ad}^c so that this extension allows observations with continuous as well as discontinuous paths.

Minimizing the cost functional (3.3) over the state-observation system (3.1)–(3.2) on the set of admissible controls U_{ad} defined by (3.4) will be called the problem (3.1)–(3.4). A control in U_{ad}, at which the functional (3.3) takes on its minimum value, is called an *optimal control.* To study the problem (3.1)–(3.4) we will use the Riccati equation

$$Q_t = \mathcal{U}^*_{T,t} Q_T \mathcal{U}_{T,t} + \int_t^T \mathcal{U}^*_{s,t}\big(F_s - (Q_s B_s + L^*_s) G_s^{-1} (B^*_s Q_s + L_s)\big)\mathcal{U}_{s,t}\, ds,\ 0 \le t \le T. \tag{3.5}$$

The conditions $(\mathbf{C_1})$–$(\mathbf{C_4})$ do not guarantee the existence of a solution of the equation (3.5) (see Curtain and Prichard [5]). Therefore, we will also need the following condition:

$(\mathbf{RE})$: The equation (3.5) has a solution in $B(\mathbf{T}, \mathcal{L}(X))$.

Note that if $Q \in B(\mathbf{T}, \mathcal{L}(X))$ is a solution of the equation (3.5), then in view of the condition $(\mathbf{C_3})$, $Q^*_t = Q_t$, $0 \le t \le T$.

Remark 3.4. The conditions $(\mathbf{C_1})$ and $(\mathbf{C_3})$ together with the condition

$(\mathbf{RE'})$: $Q_T \ge 0$, $F_t - L^*_t G_t^{-1} L_t \ge 0$ and $G_t > 0$ for a.e. $t \in \mathbf{T}$

imply the existence of a unique solution of the Riccati equation (3.5). Hence, they imply the condition $(\mathbf{RE})$ as well. The condition $(\mathbf{RE'})$ is ordinary for the control problem (3.1)–(3.4), but it is restrictive for the game considered in the next section. Therefore, we prefer the condition $(\mathbf{RE})$ rather than $(\mathbf{RE'})$.

Remark 3.5. The conditions $(\mathbf{C_1})$–$(\mathbf{C_4})$ and $(\mathbf{RE})$ do not contain anything about the relation between l, b, c, m and n. They may be independent as well as dependent. Under these conditions, observations make sense even if $C = 0$ because observing only the noise processes c and n, useful information about the state process can be obtained. Moreover, these conditions subsume the case when the observations until the current time moment t depend on the "future" of the signal noise, i.e., when the σ-algebras $\sigma(z^{u,t})$ and $\sigma(b^{t+}, m_s - m_t; t < s \le T)$ are dependent.

3.2. Separation Principle

In this section it is always assumed that conditions $(\mathbf{C_1})$–$(\mathbf{C_4})$ and $(\mathbf{RE})$ hold.

Lemma 3.6. *If the functional J takes its minimum value on U^0 at $\tilde u \in U^0$, then*

$$\tilde u_t = G_t^{-1} \mathbf{E}^0_t\,(B^*_t y_t - L_t \tilde x_t - g_t),\ \text{a.e. } t \in \mathbf{T}, \tag{3.6}$$

where $\tilde x = x^{\tilde u}$, $\mathbf{E}^0_t = \mathbf{E}\,(\cdot\,|z^{0,t})$ and

$$y_t = -\mathcal{U}^*_{T,t}(Q_T \tilde x_T + q) - \int_t^T \mathcal{U}^*_{s,t}(F_s \tilde x_s + L^*_s \tilde u_s + f_s)\, ds,\ 0 \le t \le T. \tag{3.7}$$

Proof. Let $u \in U^0$. Since U^0 is a Hilbert space, we have $\tilde u + \lambda u \in U^0$ for all $\lambda \in \mathbb{R}$. Denote

$$h_t = \int_0^t \mathcal{U}_{t,s} B_s u_s\, ds,\ 0 \le t \le T.$$

One can verify that

$$
\begin{aligned}
0 &\le J(\tilde u + \lambda u) - J(\tilde u) \\
&= \lambda^2 \mathbf{E}\langle h_T, Q_T h_T\rangle + \lambda^2 \mathbf{E}\int_0^T \big\langle h_t, (F_t - L_t^* G_t^{-1} L_t) h_t\big\rangle\, dt \\
&\quad + \lambda^2 \mathbf{E}\int_0^T \big\langle u_t + G_t^{-1} L_t h_t, G_t\big(u_t + G_t^{-1} L_t h_t\big)\big\rangle\, dt \\
&\quad + 2\lambda \mathbf{E}\int_0^T \langle G_t \tilde u_t - B_t^* y_t + L_t \tilde x_t + g_t, u_t\rangle\, dt.
\end{aligned}
$$

Dividing both sides of the above inequality by $\lambda > 0$ and then tending λ to 0, we obtain

$$
\mathbf{E}\int_0^T \langle G_t \tilde u_t - B_t^* y_t + L_t \tilde x_t + g_t, u_t\rangle\, dt \ge 0.
$$

Similarly, dividing both sides of the same inequality by $\lambda < 0$ and then tending λ to 0, we obtain

$$
\mathbf{E}\int_0^T \langle G_t \tilde u_t - B_t^* y_t + L_t \tilde x_t + g_t, u_t\rangle\, dt \le 0.
$$

Thus, since u is arbitrary in U^0, the equality (3.6) holds. □

Now let $\tilde u \in U_{\mathrm{ad}}$ be an optimal control in the problem (3.1)–(3.4). By Proposition 3.1(d), the functional J takes its minimum value on U^0 at control $\tilde u \in U_{\mathrm{ad}} \subset U^0$. Hence, by Lemma 3.6, the equalities (3.6) and (3.7) hold. Let $0 < \tau \le t \le T$. Substituting (3.6) in (3.1) and in (3.7), we have

$$
\begin{aligned}
\tilde x_t &= \mathcal{U}_{t,\tau} \tilde x_\tau + \int_\tau^t \mathcal{U}_{t,s}\, dm_s \\
&\quad + \int_\tau^t \mathcal{U}_{t,s}\big(B_s G_s^{-1} \mathbf{E}_s^0 (B_s^* y_s - L_s \tilde x_s - g_s) + b_s\big)\, ds, \\
y_t &= -\mathcal{U}_{T,t}^* (Q_T \tilde x_T + q) \\
&\quad - \int_t^T \mathcal{U}_{s,t}^* \big(F_s \tilde x_s + L_s^* G_s^{-1} \mathbf{E}_s^0 (B_s^* y_s - L_s \tilde x_s - g_s) + f_s\big)\, ds.
\end{aligned}
$$

Since $\mathbf{E}^0_\tau\,\mathbf{E}^0_s\,\xi = \mathbf{E}^0_\tau\,\xi$ for $0 < \tau \le s \le T$, we obtain

$$\begin{aligned}
\mathbf{E}^0_\tau\,\tilde{x}_t &= \mathbf{E}^0_\tau\Big(\mathcal{U}_{t,\tau}\tilde{x}_\tau + \int_\tau^t \mathcal{U}_{t,s}\,dm_s \\
&\quad + \int_\tau^t \mathcal{U}_{t,s}\left(B_sG_s^{-1}\mathbf{E}^0_s(B_s^*y_s - L_s\tilde{x}_s - g_s) + b_s\right)ds\Big), \\
&= \mathbf{E}^0_\tau\Big(\mathcal{U}_{t,\tau}\tilde{x}_\tau + \int_\tau^t \mathcal{U}_{t,s}\,dm_s \\
&\quad + \int_\tau^t \mathcal{U}_{t,s}\left(B_sG_s^{-1}(B_s^*y_s - L_s\tilde{x}_s - g_s) + b_s\right)ds\Big),
\end{aligned} \tag{3.8}$$

$$\begin{aligned}
\mathbf{E}^0_\tau\,y_t &= -\,\mathbf{E}^0_\tau\Big(\mathcal{U}^*_{T,t}(Q_T\tilde{x}_T + q) \\
&\quad + \int_t^T \mathcal{U}^*_{s,t}\big(F_s\tilde{x}_s + L_s^*G_s^{-1}\mathbf{E}^0_s(B_s^*y_s - L_s\tilde{x}_s - g_s) + f_s\big)ds\Big) \\
&= -\,\mathbf{E}^0_\tau\Big(\mathcal{U}^*_{T,t}(Q_T\tilde{x}_T + q) \\
&\quad + \int_t^T \mathcal{U}^*_{s,t}\big(F_s\tilde{x}_s + L_s^*G_s^{-1}(B_s^*y_s - L_s\tilde{x}_s - g_s) + f_s\big)ds\Big).
\end{aligned} \tag{3.9}$$

These expressions for $\mathbf{E}^0_\tau\,\tilde{x}_t$ and $\mathbf{E}^0_\tau\,y_t$ will be used in the following lemma.

Lemma 3.7. *The following equality holds:*

$$\mathbf{E}^0_\tau\,(y_t + Q_t\tilde{x}_t + \alpha_t) = 0,\ \ 0 < \tau \le t \le T, \tag{3.10}$$

where $\mathcal{Y} = \mathcal{P}_{-BG^{-1}(B^*Q+L)}(\mathcal{U})$, Q *is a solution of the equation* (3.5) *and*

$$\begin{aligned}
\alpha_t = \mathcal{Y}^*_{T,t}q &+ \int_t^T \mathcal{Y}^*_{s,t}\left(Q_sb_s - (Q_sB_s + L_s^*)G_s^{-1}g_s + f_s\right)ds \\
&+ \int_t^T \mathcal{Y}^*_{s,t}Q_s\,dm_s,\ \ 0 \le t \le T.
\end{aligned} \tag{3.11}$$

Proof. At first, we will derive an expression for $\mathbf{E}^0_\tau\,Q_t\tilde{x}_t$. For convenience, let

$$\tilde{F}_t = F_t - (Q_tB_t + L_t^*)G_t^{-1}(B_t^*Q_t + L_t),\ \ \text{a.e. } t \in \mathbf{T}.$$

Then by (3.5), we have

$$\mathbf{E}^0_\tau\,Q_t\tilde{x}_t = \mathbf{E}^0_\tau\Big(\mathcal{U}^*_{T,t}Q_T\mathcal{U}_{T,t}\tilde{x}_t + \int_t^T \mathcal{U}^*_{s,t}\tilde{F}_s\mathcal{U}_{s,t}\tilde{x}_t\,ds\Big). \tag{3.12}$$

Using (3.8) and (3.9) in (3.12), we obtain

$$\begin{aligned}
\mathbf{E}^0_\tau Q_t\tilde{x}_t &= \mathbf{E}^0_\tau \mathcal{U}^*_{T,t} Q_T \tilde{x}_T - \mathbf{E}^0_\tau \mathcal{U}^*_{T,t} Q_T \int_t^T \mathcal{U}_{T,r}\, dm_r \\
&\quad - \mathbf{E}^0_\tau \mathcal{U}^*_{T,t} Q_T \int_t^T \mathcal{U}_{T,r} \big(B_r G_r^{-1}(B_r^* y_r - L_r \tilde{x}_r - g_r) + b_r\big) dr \\
&\quad + \mathbf{E}^0_\tau \int_t^T \mathcal{U}^*_{s,t} \tilde{F}_s \tilde{x}_s\, ds - \mathbf{E}^0_\tau \int_t^T \int_t^s \mathcal{U}^*_{s,t} \tilde{F}_s \mathcal{U}_{s,r}\, dm_r\, ds \\
&\quad - \mathbf{E}^0_\tau \int_0^T \int_t^s \mathcal{U}^*_{s,t} \tilde{F}_s \mathcal{U}_{s,r} \big(B_r G_r^{-1}(B_r^* y_r - L_r \tilde{x}_r - g_r) + b_r\big) dr\, ds \\
&= \mathbf{E}^0_\tau \mathcal{U}^*_{T,t} Q_T \tilde{x}_T + \mathbf{E}^0_\tau \int_t^T \mathcal{U}^*_{s,t} \tilde{F}_s \tilde{x}_s\, ds \\
&\quad - \mathbf{E}^0_\tau \int_t^T \mathcal{U}^*_{r,t} \left(\mathcal{U}^*_{T,r} Q_T \mathcal{U}_{T,r} + \int_r^T \mathcal{U}^*_{s,r} \tilde{F}_s \mathcal{U}_{s,r}\, ds \right) dm_r \\
&\quad - \mathbf{E}^0_\tau \int_t^T \mathcal{U}^*_{r,t} \left(\mathcal{U}^*_{T,r} Q_T \mathcal{U}_{T,r} + \int_r^T \mathcal{U}^*_{s,r} \tilde{F}_s \mathcal{U}_{s,r}\, ds \right) \\
&\qquad\qquad \times \big(B_r G_r^{-1}(B_r^* y_r - L_r \tilde{x}_r - g_r) + b_r\big) dr \\
&= \mathbf{E}^0_\tau \mathcal{U}^*_{T,t} Q_T \tilde{x}_T - \mathbf{E}^0_\tau \int_t^T \mathcal{U}^*_{s,t} Q_s\, dm_s \\
&\quad + \mathbf{E}^0_\tau \int_t^T \mathcal{U}^*_{s,t} \big((F_s - (Q_s B_s + L_s^*) G_s^{-1} B_s^* Q_s - L_s^* G_s^{-1} L_s) \tilde{x}_s \\
&\qquad\qquad - Q_s B_s G_s^{-1} B_s^* y_s + Q_s B_s G_s^{-1} g_s - Q_s b_s \big) ds. \qquad (3.13)
\end{aligned}$$

Also, since $\mathcal{Y} = \mathcal{P}_{-BG^{-1}(B^*Q+L)}(\mathcal{U})$, for the random variable α defined by (3.11), we have

$$\begin{aligned}
\alpha_t = \mathcal{U}^*_{T,t} q + \int_t^T \mathcal{U}^*_{s,t} Q_s\, dm_s + \int_t^T \mathcal{U}^*_{s,t} \big(Q_s b_s - (Q_s B_s + L_s^*) G_s^{-1} g_s \\
+ f_s - (Q_s B_s + L_s^*) G_s^{-1} B_s^* \alpha_s \big) ds. \qquad (3.14)
\end{aligned}$$

Thus, from (3.9), (3.13) and (3.14), we obtain

$$\mathbf{E}^0_\tau (y_t + Q_t \tilde{x}_t + \alpha_t) = \int_t^T \mathcal{U}^*_{s,t} (Q_s B_s + L_s^*) G_s^{-1} B_s^* \, \mathbf{E}^0_\tau (y_s + Q_s \tilde{x}_s + \alpha_s)\, ds.$$

If $\lambda_{t,\tau} = \big\| \mathbf{E}^0_\tau (y_t + Q_t \tilde{x}_t + \alpha_t) \big\|$, then

$$\lambda_{t,\tau} \leq c \int_t^T \lambda_{s,\tau}\, ds, \;\; c = \text{const.} \geq 0,$$

which by Gronwall's theorem implies $\lambda_{t,\tau} = 0$, $0 < \tau \leq t \leq T$. □

Theorem 3.8. (The Extended Separation Principle) *Under conditions* $(\mathbf{C_1})$–$(\mathbf{C_4})$ *and* $(\mathbf{RE})$, *let* $\tilde{u} \in U_{\text{ad}}$ *be an optimal control in the problem* (3.1)–(3.4). *Denote*

$$\tilde{x} = x^{\tilde{u}}, \ \tilde{\mathbf{E}}_t = \mathbf{E}\left(\cdot \,|z^{\tilde{u},t}\right), \ 0 < t \le T.$$

Then $\tilde{u}$ *has the form*

$$\tilde{u}_t = -\, G_t^{-1}(B_t^* Q_t + L_t)\, \tilde{\mathbf{E}}_t \tilde{x}_t - G_t^{-1} B_t^*\, \tilde{\mathbf{E}}_t\, \alpha_t - G_t^{-1}\, \tilde{\mathbf{E}}_t\, g_t, \ \text{a.e. } t \in \mathbf{T}, \tag{3.15}$$

where $\mathcal{Y} = \mathcal{P}_{-BG^{-1}(B^*Q+L)}(\mathcal{U})$, Q *is a solution of the equation* (3.5) *and* α *is defined by* (3.11). *Moreover, this optimal control is unique if equation* (3.5) *has a unique solution.*

Proof. Substituting $\tau = t$ in (3.10), we have

$$\mathbf{E}_t^0\, y_t = -\, \mathbf{E}_t^0\, (Q_t \tilde{x}_t + \alpha_t), \ 0 < t \le T.$$

Since $\tilde{u} \in U_{\text{ad}}$, by Proposition 3.1(a), $U_t^{\tilde{u}} = U_t^0$ for all $0 < t \le T$. This implies $\tilde{\mathbf{E}}_t = \mathbf{E}_t^0$ for all $0 < t \le T$. So, from (3.6), we obtain (3.15). Obviously, (3.15) uniquely determines the optimal control (if it exists) if equation (3.5) has a unique solution. □

Theorem 3.8 expresses the separation principle when the relation between the noise processes acting to state and to observations is arbitrary. By (3.15), synthesizing an optimal control is separated into two steps, the first of them being computing the conditional expectations $\tilde{\mathbf{E}}_t\, \tilde{x}_t$, $\tilde{\mathbf{E}}_t\, \alpha_t$, $\tilde{\mathbf{E}}_t\, g_t$ and the second finding an optimal control by a deterministic way using the results of the first step.

By (3.15) and (3.11), the optimal control at the time moment t depends on random processes b and m on $(t, T]$. This is natural since we consider arbitrarily related noise processes and, therefore, the observations available up to the time moment t may contain information about the "future" of the signal noise, i.e., the σ-algebras $\sigma\big(z_t^{\tilde{u},t}\big)$ and $\sigma\big(b^{t+}, m_s - m_t; t < s \le T\big)$ may be dependent (see Remark 3.5). The formulae (3.15) and (3.11) show how this information must be used in the optimal control.

It is natural to call the statement of Theorem 3.8 the *extended separation principle* and the following particular case of Theorem 3.8 the *classic separation principle.*

Theorem 3.9. (The Classic Separation Principle) *Under conditions* $(\mathbf{C_1})$–$(\mathbf{C_3})$, $(\mathbf{RE})$ *and*

- $(\mathbf{C_4'})$ $q \in X$, $f \in L_2(\mathbf{T}, X)$, $g \in L_2(\mathbf{T}, U)$,
- $(\mathbf{C_5})$ $\sigma(l, b^t, c^t, n_s, m_s; 0 \le s \le t)$ *and* $\sigma(b^{t+}, m_s - m_t; t < s \le T)$ *are independent for all* $0 < t \le T$,

let $\tilde{u} \in U_{\text{ad}}$ *be an optimal control in the problem* (3.1)–(3.4). *Denote*

$$\tilde{x} = x^{\tilde{u}}, \ \tilde{\mathbf{E}}_t = \mathbf{E}\left(\cdot \,|z^{\tilde{u},t}\right), \ 0 < t \le T.$$

Then $\tilde{u}$ *has the form*

$$\tilde{u}_t = u_t^0 - G_t^{-1}(B_t^* Q_t + L_t)\, \tilde{\mathbf{E}}_t\, \tilde{x}_t, \ \text{a.e. } t \in \mathbf{T},$$

where $\mathcal{Y} = \mathcal{P}_{-BG^{-1}(B^*Q+L)}(\mathcal{U})$, Q *is a solution of equation* (3.5) *and* u^0 *is a function in* $L_2(\mathbf{T}, U)$ *defined by*

$$u_t^0 = -G_t^{-1}g_t - G_t^{-1}B_t^*\mathcal{Y}_{T,t}^*q - G_t^{-1}B_t^*\int_t^T \mathcal{Y}_{s,t}^*(Q_s\mathbf{E}\,b_s - (Q_sB_s + L_s^*)G_s^{-1}g_s + f_s)ds, \text{ a.e. } t \in \mathbf{T}.$$

Moreover, this optimal control is unique if equation (3.5) *has a unique solution.*

Proof. By ($\mathbf{C_6}$), we have

$$\tilde{\mathbf{E}}_t\left(\int_t^T \mathcal{Y}_{s,t}^*Q_sb_s\,ds + \int_t^T \mathcal{Y}_{s,t}^*Q_s\,dm_s\right) = \int_t^T \mathcal{Y}_{s,t}^*Q_s\,\mathbf{E}\,b_s\,ds.$$

So, the statement follows from Theorem 3.8. □

Comparing Theorems 3.8 and 3.9, one can observe that Theorem 3.8 removes the only essential restriction (that is the condition ($\mathbf{C_5}$)) under which the classic separation principle is true. This weakening completes the form of the respective optimal control with the additional terms.

3.3. Existence of Optimal Control

Theorems 3.8 and 3.9 determine the form of the optimal control under its existence. In the case of the classic separation principle it is known that the optimal control exists if there exists a Kalman–Bucy type optimal filter in the associated filtering problem. In this section this fact is established for the extended separation principle.

Consider the problem (3.1)–(3.4) and assume that conditions ($\mathbf{C_1}$)–($\mathbf{C_3}$), ($\mathbf{C_4'}$) and ($\mathbf{RE}'$) hold and n is a Wiener martingale, i.e.,

($\mathbf{E_1}$) n has the representation

$$n_t = \int_0^t \Psi_s\,dv_s,\ 0 \le t \le T,$$

where $\Psi \in B_\infty(\mathbf{T}, \mathcal{L}(Y, Z))$ and v is a Y-valued Wiener process.

A function in the form

$$u_t = u_t^0 + \int_0^t M_{t,s}\,dz_s^u, \text{ a.e. } t \in \mathbf{T}, \tag{3.16}$$

where $u^0 \in L_2(\mathbf{T}, U)$ and $M \in B_2(\Delta_T, \mathcal{L}(Z, U))$, is called a *control in linear feedback form*. Note that by ($\mathbf{E_1}$), the integral in (3.16) is well defined.

Lemma 3.10. *If* u *has the representation* (3.16) *with* $u^0 \in L_2(\mathbf{T}, U)$ *and with* $M \in B_2(\Delta_T, \mathcal{L}(Z, U))$, *then it also has the representation*

$$u_t = u_t^1 + \int_0^t N_{t,s}\,dz_s^0, \text{ a.e. } t \in \mathbf{T}, \tag{3.17}$$

for some $u^1 \in L_2(\mathbf{T}, U)$ *and* $N \in B_2(\Delta_T, \mathcal{L}(Z, U))$ *and vice versa, where* z^0 *is the observation process* (3.2) *corresponding to zero control.*

Proof. Suppose (3.16) holds. By (3.1)–(3.2), we have

$$z_t^u = z_t^0 + \int_0^t \int_0^s C_s \mathcal{U}_{s,r} B_r u_r \, dr \, ds, \; 0 \le t \le T. \tag{3.18}$$

Using (3.18) in (3.16), we obtain

$$u_t = u_t^0 + \int_0^t M_{t,s} \, dz_s^0 + \int_0^t \int_0^s M_{t,s} C_s \mathcal{U}_{s,r} B_r u_r \, dr \, ds, \; \text{a.e. } t \in \mathbf{T}. \tag{3.19}$$

The equation (3.19) is a Volterra integral equation in u. It is well known that there exists $G \in B_2(\Delta_T, \mathcal{L}(U))$ satisfying

$$u_t = u_t^0 + \int_0^t M_{t,s} \, dz_s^0 + \int_0^t G_{t,s} \left(u_s^0 + \int_0^s M_{s,r} \, dz_r^0 \right) ds, \; \text{a.e. } t \in \mathbf{T}.$$

Hence, for

$$u_t^1 = u_t^0 + \int_0^t G_{t,s} u_s^0 \, ds, \; \text{a.e. } t \in \mathbf{T},$$

and for

$$N_{t,s} = M_{t,s} + \int_s^t G_{t,r} M_{r,s} \, dr, \; 0 \le s \le t \le T,$$

we obtain the representation (3.17) for u. In a similar manner, starting from (3.17), the representation (3.16) can be obtained. □

Proposition 3.11. *The set U_{ad}, defined by* (3.4), *contains all controls that are in the linear feedback form* (3.16) *with*

$$u^0 \in L_2(\mathbf{T}, U) \text{ and } M \in B_2(\Delta_T, \mathcal{L}(Z, U)).$$

Proof. If u has the form (3.16), then $u \in U^1$. By Lemma 3.10, u has the form (3.17) as well. Hence, $u \in U^0$. Thus, $u \in U^0 \cap U^1 = U_{\mathrm{ad}}$. □

Proposition 3.12. *The set U_{ad}, defined by* (3.4), *contains all controls in the form* (3.17) *where $u^1 \in L_2(\mathbf{T}, U)$ and $N \in B_2(\Delta_T, \mathcal{L}(Z, U))$.*

Proof. This is similar to the proof of Proposition 3.11. □

Under conditions $(\mathbf{C_1})$–$(\mathbf{C_3})$, $(\mathbf{C_4'})$ and $(\mathbf{RE'})$, the functional (3.3) is strictly convex on the Hilbert space U^0 and, hence, it takes its minimum value on U^0 at the unique optimal control $\tilde{u}$ which, by Lemmas 3.6 and 3.7, has the form

$$\tilde{u}_t = -G_t^{-1}(B_t^* Q_t + L_t)\mathbf{E}_t^0 \tilde{x}_t - G_t^{-1} B_t^* \mathbf{E}_t^0 \alpha_t - G_t^{-1} g_t, \; \text{a.e. } t \in \mathbf{T}. \tag{3.20}$$

Let us find another representation for this $\tilde{u}$. Denote

$$x_t = \mathcal{U}_{t,0}(l - \mathbf{E}l) + \int_0^t \mathcal{U}_{t,s}(b_s - \mathbf{E}b_s) \, ds + \int_0^t \mathcal{U}_{t,s} \, dm_s, \; 0 \le t \le T, \tag{3.21}$$

$$\eta_t = \mathcal{U}_{t,0}\mathbf{E}l + \int_0^t \mathcal{U}_{t,s}(B_s \tilde{u}_s + \mathbf{E}b_s) \, ds, \; 0 \le t \le T. \tag{3.22}$$

Then

$$\mathbf{E}_t^0 \tilde{x}_t = \mathbf{E}_t^0 x_t + \eta_t, \ 0 < t \leq T. \tag{3.23}$$

Using (3.23) in (3.20), we have

$$\tilde{u} = -G_t^{-1}(B_t^* Q_t + L_t)\big(\mathbf{E}_t^0 x_t + \eta_t\big) - G_t^{-1} B_t^* \mathbf{E}_t^0 \alpha_t - G_t^{-1} g_t, \text{ a.e. } t \in \mathbf{T}. \tag{3.24}$$

From $\mathcal{Y} = \mathcal{P}_{-BG^{-1}(B^*Q+L)}(\mathcal{U})$, we obtain

$$\begin{aligned} \eta_t &= \mathcal{U}_{t,0}\mathbf{E}l \\ &\quad + \int_0^t \mathcal{U}_{t,s}\big(\mathbf{E}b_s - B_s G_s^{-1}\big((B_s^* Q_s + L_s)(\mathbf{E}_t^0 x_t + \eta_s) - g_s - B_s^* \mathbf{E}_s^0 \alpha_s\big)\big) ds \\ &= \mathcal{Y}_{t,0}\mathbf{E}l \\ &\quad + \int_0^t \mathcal{Y}_{t,s}\big(\mathbf{E}b_s - B_s G_s^{-1}\big((B_s^* Q_s + L_s)\mathbf{E}_t^0 x_t - g_s - B_s^* \mathbf{E}_s^0 \alpha_s\big)\big) ds. \end{aligned} \tag{3.25}$$

Substituting (3.25) and (3.11) in (3.24), we have

$$\begin{aligned} \tilde{u}_t &= u_t^1 - G_t^{-1}\mathbf{E}_t^0 y_t \\ &\quad + G_t^{-1}(B_t^* Q_t + L_t)\int_0^t \mathcal{Y}_{t,s} B_s G_s^{-1} \mathbf{E}_s^0 y_s \, ds, \text{ a.e. } t \in \mathbf{T}, \end{aligned} \tag{3.26}$$

whcre

$$\begin{aligned} y_t &= (B_t^* Q_t + L_t) x_t + B_t^* \int_t^T \mathcal{Y}_{s,t}^* Q_s (b_s - \mathbf{E}b_s) \, ds \\ &\quad + B_t^* \int_t^T \mathcal{Y}_{s,t}^* Q_s \, dm_s, \ 0 \leq t \leq T, \end{aligned} \tag{3.27}$$

$$\begin{aligned} u_t^1 &= -G_t^{-1} g_t - G_t^{-1} B_t^* \mathcal{Y}_{T,t}^* q - G_t^{-1}(B_t^* Q_t + L_t)\mathcal{Y}_{t,0}\mathbf{E}l \\ &\quad + G_t^{-1} B_t^* \int_t^T \mathcal{Y}_{s,t}^* \big((Q_s B_s + L_s^*) G_s^{-1} g_s - Q_s \mathbf{E}b_s - f_s\big) ds \\ &\quad - G_t^{-1}(B_t^* Q_t + L_t)\int_0^t \mathcal{Y}_{t,s}\big(\mathbf{E}b_s - B_s G_s^{-1}(g_s + B_s^* \mathcal{Y}_{T,s} q)\big) ds \\ &\quad - G_t^{-1}(B_t^* Q_t + L_t)\int_0^t \int_s^T \mathcal{Y}_{t,s} B_s G_s^{-1} B_s^* \mathcal{Y}_{r,s}^* \big((Q_r B_r + L_r^*) G_r^{-1} g_r \\ &\qquad\qquad - Q_r \mathbf{E}b_r - f_r\big) dr \, ds, \text{ a.e. } t \in \mathbf{T}. \end{aligned} \tag{3.28}$$

Let

$$z_t = \int_0^t (C_s x_s + c_s - \mathbf{E}c_s) \, ds + \int_0^t \Psi_s \, dv_s, \ 0 \leq t \leq T, \tag{3.29}$$

where x is defined by (3.21). Obviously, the difference $z_t^0 - z_t$ is nonrandom for all $0 \leq t \leq T$. Also, by $(\mathbf{E_1})$, z and z^0 have continuous paths. So, we can write

$$\mathbf{E}_t^0 y_t = \mathbf{E}(y_t | z_s^0; 0 < s \leq t) = \mathbf{E}(y_t | z_s; 0 < s \leq t), \ 0 < t \leq T. \tag{3.30}$$

We will refer to the condition:

($\mathbf{E_2}$) For y and z, defined by (3.21), (3.27) and (3.29),

$$\mathbf{E}(y_t|z_s; 0 < s \le t) = \int_0^t K_{t,s}\, dz_s,\ 0 < t \le T,$$

where $K \in B_2(\Delta_T, \mathcal{L}(Z, U))$.

Theorem 3.13. *Under conditions* ($\mathbf{C_1}$)–($\mathbf{C_3}$), ($\mathbf{C'_4}$), ($\mathbf{RE'}$) *and* ($\mathbf{E_1}$)–($\mathbf{E_2}$), *there exists a unique optimal control in the problem* (3.1)–(3.4).

Proof. By (3.30) and ($\mathbf{E_2}$), the formula (3.26) defines a control $\tilde{u}$ in the form (3.17) for some $u^1 \in L_2(\mathbf{T}, U)$ and $N \in B_2(\Delta_T, \mathcal{L}(Z, U))$. Therefore, by Proposition 3.12, $\tilde{u}$ belongs to U_{ad}. On the other hand, by the conditions ($\mathbf{C_3}$) and ($\mathbf{RE'}$), the functional (3.3) is strictly convex on U^0. Hence, it takes its minimum value on U^0 at $\tilde{u}$, defined by (3.20), which is equal to $\tilde{u}$, defined by (3.26). Therefore, by Proposition 3.1(d), $\tilde{u}$ is an optimal control in the problem (3.1)–(3.4). The uniqueness of the optimal control follows from Theorem 3.8 and Remark 3.4. □

Now let

$$x_t = \mathcal{U}_{t,0}x_0 + \int_0^t \mathcal{U}_{t,s}\varphi_s^1\, ds + \int_0^t \mathcal{U}_{t,s}\, dm_s,\ 0 \le t \le T, \tag{3.31}$$

$$y_t = M_t x_t + \int_t^T N_{s,t}\varphi_s^1\, ds + \int_t^T N_{s,t}\, dm_s,\ 0 \le t \le T, \tag{3.32}$$

$$z_t = \int_0^t (C_s x_s + \varphi_s^2)\, ds + \int_0^t \Psi_s\, dv_s,\ 0 \le t \le T. \tag{3.33}$$

One can observe that the random processes x, y and z defined by (3.31)–(3.33) are the same as x, y and z defined by the formulae (3.21), (3.27) and (3.29) if

$$x_0 = l - \mathbf{E}l,\ \varphi_t^1 = b_t - \mathbf{E}b_t,\ \varphi_t^2 = c_t - \mathbf{E}c_t,\ 0 \le t \le T,$$
$$N_{s,t} = B_t^* \mathcal{Y}_{s,t}^* Q_s,\ (s,t) \in \Delta_T,\ M_t = B_t^* Q_t + L_t,\ 0 \le t \le T.$$

Indeed condition ($\mathbf{E_2}$) from Section 3.3, which is essential in Theorem 3.13, means finding the best estimate of y_t based on the observations z_s, $0 \le s \le t$, in the linear feedback form for all $0 < t \le T$. With the notation introduced above, let us call this problem the filtering problem (3.31)–(3.33).

Two cases in the filtering problem (3.31)–(3.33) must be specified. The first case assumes that the past of observations and the "future" of signal noise are independent, i.e., $\sigma(z_s; 0 \le s \le t)$ and $\sigma(\varphi_s^1, m_s - m_t; t < s \le T)$ are independent for all $0 < t \le T$. In this case,

$$\mathbf{E}_t y_t = \mathbf{E}(y_t|z_s; 0 \le s \le t) = M_t \mathbf{E}(x_t|z_s; 0 \le s \le t),\ 0 < t \le T,$$

and the filtering problem (3.31)–(3.33) reduces to estimating x_t based on the observations z_s, $0 \le s \le t$. This case includes the classic Kalman–Bucy filtering under independent or correlated white noise processes as well.

The second case is more complicated and requires the dependence of the σ-algebras $\sigma(z_s; 0 \le s \le t)$ and $\sigma(\varphi_s^1, m_s - m_t; t < s \le T)$ for the values of t taken from some nontrivial subinterval of $\mathbf{T}$. In particular, this case includes filtering under coloured noise processes, widely represented in the existing literature (see, for example, Bucy and Joseph [8]). In another way this case can be achieved when the noise processes of the signal system (3.31) and of the observation system (3.33) are distributed or pointwise delays of each other. Note that the distributed delays of white noise processes lead to wide band noise processes. The filtering problems under delayed noise processes are discussed in Bashirov [9] and Bashirov and Mishne [10].

4. Game Problem

4.1. Setting of the Problem

Consider the state system (3.1), in which the Hilbert space U is decomposed into the product of two Hilbert spaces U_1 and U_2. In this case, each function $u \in L_2(\mathbf{T} \times \Omega, U)$ can be represented as a pair of $u^1 \in L_2(\mathbf{T} \times \Omega, U_1)$ and $u^2 \in L_2(\mathbf{T} \times \Omega, U_2)$. The functions u_1 and u_2 will be considered as control actions of the first and second players, respectively. In view of decomposition $U = U_1 \times U_2$, the functions B, L, G and q from (3.1) and (3.3) could be decomposed in a respective way. Also, it will be supposed that each player has his own observation system in the form (3.2). Realizing the above, we obtain the state-observation system

$$x_t^{u^1,u^2} = \mathcal{U}_{t,0} l + \int_0^t \mathcal{U}_{t,s}\left([B_{1,s} \ B_{2,s}] \begin{bmatrix} u_s^1 \\ u_s^2 \end{bmatrix} + b_s \right) ds + \int_0^t \mathcal{U}_{t,s}\, dm_s, \ 0 \le t \le T, \tag{4.1}$$

$$z_{i,t}^{u^1,u^2} = \int_0^t \left(C_{i,s} x_s^{u^1,u^2} + c_s^i \right) ds + n_t^i, \ 0 \le t \le T, \ i = 1, 2, \tag{4.2}$$

and the quadratic functional

$$\begin{aligned} J(u^1, u^2) = \mathbf{E}\Bigg(& \left\langle x_T^{u^1,u^2}, Q_T x_T^{u^1,u^2} \right\rangle \\ & + \int_0^T \left\langle \begin{bmatrix} x_t^{u^1,u^2} \\ u_t^1 \\ u_t^2 \end{bmatrix}, \begin{bmatrix} F_t & L_{1,t}^* & L_{2,t}^* \\ L_{1,t} & G_{1,t} & R_t^* \\ L_{2,t} & R_t & G_{2,t} \end{bmatrix} \begin{bmatrix} x_t^{u^1,u^2} \\ u_t^1 \\ u_t^2 \end{bmatrix} \right\rangle dt \\ & + 2\left\langle q, x_T^{u^1,u^2} \right\rangle + 2 \int_0^T \left\langle \begin{bmatrix} f_t \\ g_t^1 \\ g_t^2 \end{bmatrix}, \begin{bmatrix} x_t^{u^1,u^2} \\ u_t^1 \\ u_t^2 \end{bmatrix} \right\rangle dt \Bigg). \end{aligned} \tag{4.3}$$

In this section we suppose that the following conditions hold:

- **(G₁)** $\mathcal{U} \in \mathcal{E}(\Delta_T, \mathcal{L}(X))$, $B_i \in B_\infty(\mathbf{T}, \mathcal{L}(U_i, X))$, $C_i \in B_\infty(\mathbf{T}, \mathcal{L}(X, Z_i))$, $i = 1, 2$;
- **(G₂)** $l \in L_2(\Omega, X)$, $b \in L_2(\mathbf{T} \times \Omega, X)$, $c^i \in L_2(\mathbf{T} \times \Omega, Z_i)$, $m \in M_2(\mathbf{T}, X)$, $n^i \in M_2(\mathbf{T}, Z_i)$, $i = 1, 2$;

($\mathbf{G_3}$) $Q_T \in \mathcal{L}(X)$, $Q_T^* = Q_T$, $F \in B_\infty(\mathbf{T}, \mathcal{L}(X))$, $G_i, G_i^{-1} \in B_\infty(\mathbf{T}, \mathcal{L}(U_i))$, $R \in B_\infty(\mathbf{T}, \mathcal{L}(U_1, U_2))$, $F_t^* = F_t$, $G_{1,t} > 0$ and $G_{2,t} < 0$ for a.e. $t \in \mathbf{T}$, $L_i \in B_\infty(\mathbf{T}, \mathcal{L}(X, U_i))$, $(G_1 - R^* G_2^{-1} R)^{-1} \in B_\infty(\mathbf{T}, \mathcal{L}(U_1))$, $(G_2 - R G_1^{-1} R^*)^{-1} \in B_\infty(\mathbf{T}, \mathcal{L}(U_2))$, $i = 1, 2$;

($\mathbf{G_4}$) $q \in L_2(\Omega, X)$, $f \in L_2(\mathbf{T} \times \Omega, X)$, $g^i \in L_2(\mathbf{T} \times \Omega, U_i)$, $i = 1, 2$.

As mentioned above, we denote $U = U_1 \times U_2$ and

$$u = \begin{bmatrix} u^1 \\ u^2 \end{bmatrix}, \; B = [B_1 \; B_2], \; L = \begin{bmatrix} L^1 \\ L^2 \end{bmatrix}, \; G = \begin{bmatrix} G_1 & R^* \\ R & G_2 \end{bmatrix}, \; g = \begin{bmatrix} g^1 \\ g^2 \end{bmatrix}.$$

We will need the following proposition.

Proposition 4.1. *Let*

$$S = \begin{bmatrix} N_1 & M^* \\ M & N_2 \end{bmatrix},$$

where $N_1 \in \mathcal{L}(U_1)$, $N_2 \in \mathcal{L}(U_2)$, $M \in \mathcal{L}(U_1, U_2)$, $N_1 > 0$ *and* $N_2 < 0$. *Then there exists* $S^{-1} \in \mathcal{L}(U_1 \times U_2)$ *and* $S^{-1} = S_1 = S_2$, *where*

$$S_1 = \begin{bmatrix} \left(N_1 - M^* N_2^{-1} M\right)^{-1} & -N_1^{-1} M^* \left(N_2 - M N_1^{-1} M^*\right)^{-1} \\ -N_2^{-1} M \left(N_1 - M^* N_2^{-1} M\right)^{-1} & \left(N_2 - M N_1^{-1} M^*\right)^{-1} \end{bmatrix},$$

$$S_2 = \begin{bmatrix} \left(N_1 - M^* N_2^{-1} M\right)^{-1} & -\left(N_1 - M^* N_2^{-1} M\right)^{-1} M^* N_2^{-1} \\ -\left(N_2 - M N_1^{-1} M^*\right)^{-1} M N_1^{-1} & \left(N_2 - M N_1^{-1} M^*\right)^{-1} \end{bmatrix}.$$

Proof. First, note that $N_1 > 0$ and $N_2 < 0$ imply $N_1 - M^* N_2^{-1} M > 0$ and, therefore, there exists $(N_1 - M^* N_2^{-1} M)^{-1} \in \mathcal{L}(U_1)$. Similarly, there exists $(N_2 - M N_1^{-1} M^*)^{-1} \in \mathcal{L}(U_2)$. So, $S_1, S_2 \in \mathcal{L}(U_1 \times U_2)$. Let

$$v = \begin{bmatrix} v_1 \\ v_2 \end{bmatrix} \in U_1 \times U_2 \text{ and } Sv = w = \begin{bmatrix} w_1 \\ w_2 \end{bmatrix} \in U_1 \times U_2.$$

We have

$$\begin{aligned} Sv = w &\Leftrightarrow \left\{ \begin{array}{l} N_1 v_1 + M^* v_2 = w_1 \\ M v_1 + N_2 v_2 = w_2 \end{array} \right\} \Leftrightarrow \left\{ \begin{array}{l} v_1 = N_1^{-1} w_1 - N_1^{-1} M^* v_2 \\ v_2 = N_2^{-1} w_2 - N_2^{-1} M v_1 \end{array} \right\} \\ &\Leftrightarrow \left\{ \begin{array}{l} N_1 v_1 + M^* N_2^{-1} w_2 - M^* N_2^{-1} M v_1 = w_1 \\ N_2 v_2 + M N_1^{-1} w_1 - M N_1^{-1} M^* v_2 = w_2 \end{array} \right\} \\ &\Leftrightarrow \left\{ \begin{array}{l} v_1 = \left(N_1 - M^* N_2^{-1} M\right)^{-1} \left(w_1 - M^* N_2^{-1} w_2\right) \\ v_2 = \left(N_2 - M N_1^{-1} M^*\right)^{-1} \left(w_2 - M N_1^{-1} w_1\right) \end{array} \right\} \Leftrightarrow v = S_2 w. \end{aligned}$$

So, S_2 is the right inverse of S, i.e., $SS_2 = I$. Since S is self-adjoint and $S_2^* = S_1$, we easily obtain that $S_1 S = I$, i.e., S_1 is the left inverse of S. In view of

$$\begin{aligned} N_1^{-1} M^* &= \left(N_1 - M^* N_2^{-1} M\right)^{-1} \left(N_1 - M^* N_2^{-1} M\right) N_1^{-1} M^* \\ &= \left(N_1 - M^* N_2^{-1} M\right)^{-1} M^* N_2^{-1} \left(N_2 - M N_1^{-1} M^*\right), \end{aligned}$$

we conclude that

$$N_1^{-1} M^* \left(N_2 - M N_1^{-1} M^*\right)^{-1} = \left(N_1 - M^* N_2^{-1} M\right)^{-1} M^* N_2^{-1}.$$

Thus, $S_1 = S_2 = S^{-1} \in \mathcal{L}(U_1 \times U_2)$. □

By Proposition 4.1, the condition $(\mathbf{G_3})$ implies the existence of G_t^{-1} for a.e. $t \in \mathbf{T}$ and $G^{-1} \in B_\infty(\mathbf{T}, \mathcal{L}(U))$. Therefore, we can consider the operator Riccati equation (3.5). We suppose that the condition (**RE**) from Section 3.1 holds.

The set of admissible controls of each player we define as in the problem (3.1)–(3.4). For $u^1 \in L_2(\mathbf{T} \times \Omega, U_1)$ and $u^2 \in L_2(\mathbf{T} \times \Omega, U_2)$, consider

$$U_{i,t}^{u^1,u^2} = L_2\left(\Omega, \sigma\left(z_i^{u^1,u^2,t}\right), \mathbf{P}, U_i\right), \ i = 1, 2, \ 0 < t \le T,$$

where $z_i^{u^1,u^2,t}$ denotes an $L_2(0, t; Z_i)$-valued random variable, obtained by restriction of $z_i^{u^1,u^2}$ from $\mathbf{T} \times \Omega$ to $[0, t] \times \Omega$. Let

$$U_1^{0,u^2} = \int_0^T U_{1,t}^{0,u^2}\, dt = \left\{u^1 \in L_2(\mathbf{T} \times \Omega, U_1) : u_t^1 \in U_{1,t}^{0,u^2} \text{ for a.e. } t \in \mathbf{T}\right\},$$

$$U_2^{u^1,0} = \int_0^T U_{2,t}^{u^1,0}\, dt = \left\{u^2 \in L_2(\mathbf{T} \times \Omega, U_2) : u_t^2 \in U_{2,t}^{u^1,0} \text{ for a.e. } t \in \mathbf{T}\right\}.$$

Also, define

$$U_1^{u^2} = \left\{u^1 \in L_2(\mathbf{T} \times \Omega, U_1) : u_t^1 \in U_{1,t}^{u^1,u^2} \text{ for a.e. } t \in \mathbf{T}\right\},$$

$$U_2^{u^1} = \left\{u^2 \in L_2(\mathbf{T} \times \Omega, U_2) : u_t^2 \in U_{2,t}^{u^1,u^2} \text{ for a.e. } t \in \mathbf{T}\right\}.$$

The set of admissible controls in the considered game will be defined as

$$U_{\text{ad}} = \left\{(u^1, u^2) : u^1 \in \left(U_1^{0,u^2} \cap U_1^{u^2}\right), u^2 \in \left(U_2^{u^1,0} \cap U_2^{u^1}\right)\right\}, \tag{4.4}$$

where u^1 and u^2 are admissible controls of the first and second players, respectively. Note that for fixed u^1, the cross section $U_2^{u^1,0} \cap U_2^{u^1}$ of U_{ad} agrees with the set of admissible controls defined for the control problem (3.1)–(3.4). The same holds for the cross sections at fixed u^2.

Our aim in this section is to study a saddle point of the functional (4.3) on U_{ad}. A *saddle point* is a pair $(\tilde{u}^1, \tilde{u}^2) \in U_{\text{ad}}$ such that

$$J(\tilde{u}^1, u^2) \le J(\tilde{u}^1, \tilde{u}^2) \le J(u^1, \tilde{u}^2) \tag{4.5}$$

for all $u^1 \in \left(U_1^{0,\tilde{u}^2} \cap U_1^{\tilde{u}^2}\right)$ and for all $u^2 \in \left(U_2^{\tilde{u}^1,0} \cap U_2^{\tilde{u}^1}\right)$. Briefly, this problem will be called the game (4.1)–(4.4). The technique used in studying the control problem (3.1)–(3.4) will be applied to study the game (4.1)–(4.4). Two cases will be considered. In the first case the observations of the first player will be included in the observations of the second player. In the second case the players will have the same observations.

4.2. Case 1: The First Player Has Worse Observations

The mathematical condition corresponding to the considered case is formulated as

$(\mathbf{G}_5')$ $\sigma\left(z_1^{u^1,u^2,t}\right) \subset \sigma\left(z_2^{u^1,u^2,t}\right)$ for all $(u^1,u^2) \in U_{\text{ad}}$ and for all $0<t\le T$.

The condition $(\mathbf{G}_5')$ means that the second player is completely informed about the observations of the first player, i.e., the first player has worse observations than the second one.

Consider the game (4.1)–(4.4) under conditions $(\mathbf{G_1})$–$(\mathbf{G_4})$, $(\mathbf{RE})$ and $(\mathbf{G}_5')$. Suppose $(\tilde u^1,\tilde u^2) \in U_{\text{ad}}$ is a saddle point in the game (4.1)–(4.4). Denote

$$\tilde u = \begin{bmatrix} \tilde u^1 \\ \tilde u^2 \end{bmatrix}, \ \tilde x = x^{\tilde u^1,\tilde u^2}, \ \mathbf{E}_{i,t}^{u^1,u^2} = \mathbf{E}\left(\cdot \left| z_i^{u^1,u^2,t}\right.\right), \ i=1,2, \ 0<t\le T,$$

and consider the random process y defined by (3.7). Since the functional (4.3) takes its minimum value at

$$\tilde u^1 \in \left(U_1^{0,\tilde u^2} \cap U_1^{\tilde u^2}\right) \subset U_1^{0,\tilde u^2}$$

under fixed $u^2=\tilde u^2$, then by Lemma 3.6,

$$\tilde u_t^1 = G_{1,t}^{-1}\mathbf{E}_{1,t}^{0,\tilde u^2}\left(B_{1,t}^* y_t - L_{1,t}\tilde x_t - g_t^1 - R_t^*\tilde u_t^2\right), \ \text{a.e. } t\in\mathbf{T}. \tag{4.6}$$

In a similar way, we have

$$\tilde u_t^2 = G_{2,t}^{-1}\mathbf{E}_{2,t}^{\tilde u^1,0}\left(B_{2,t}^* y_t - L_{2,t}\tilde x_t - g_t^2 - R_t\tilde u_t^1\right), \ \text{a.e. } t\in\mathbf{T}. \tag{4.7}$$

According to the condition $(\mathbf{G}_5')$ and Proposition 3.1(a),

$$U_{1,t}^{0,\tilde u^2} = U_{1,t}^{\tilde u^1,\tilde u^2} \subset U_{2,t}^{\tilde u^1,\tilde u^2} = U_{2,t}^{\tilde u^1,0}, \ 0<t\le T. \tag{4.8}$$

Hence,

$$\mathbf{E}_{2,t}^{\tilde u^1,0}\,\tilde u_t^1 = \mathbf{E}_{2,t}^{\tilde u^1,0}\,\mathbf{E}_{1,t}^{0,\tilde u^2}\,\tilde u_t^1 = \mathbf{E}_{1,t}^{0,\tilde u^2}\,\tilde u_t^1 = \tilde u_t^1, \ \text{a.e. } t\in\mathbf{T}.$$

So, by (4.7),

$$\tilde u_t^2 = G_{2,t}^{-1}\mathbf{E}_{2,t}^{\tilde u^1,0}\left(B_{2,t}^* y_t - L_{2,t}\tilde x_t - g_t^2\right) - G_{2,t}^{-1}R_t\tilde u_t^1, \ \text{a.e. } t\in\mathbf{T}.$$

Using this equality in (4.6), we obtain

$$\begin{aligned}\tilde u_t^1 = \left(G_{1,t} - R_t^* G_{2,t}^{-1}R_t\right)^{-1}\mathbf{E}_{1,t}^{0,\tilde u^2}\Big(&\left(B_{1,t}^* - R_t^*G_{2,t}^{-1}B_{2,t}^*\right)y_t \\ &- \left(L_{1,t} - R_t^*G_{2,t}^{-1}L_{2,t}\right)\tilde x_t - g_t^1 + R_t^*G_{2,t}^{-1}g_t^2\Big), \ \text{a.e. } t\in\mathbf{T}.\end{aligned} \tag{4.9}$$

Now let $0<\tau\le t\le T$. Using (4.8) in the formulae (4.6) and (4.7), we have

$$\mathbf{E}_{1,\tau}^{0,\tilde u^2}\,\tilde u_t^1 = G_{1,t}^{-1}\mathbf{E}_{1,\tau}^{0,\tilde u^2}\left(B_{1,t}^* y_t - L_{1,t}\tilde x_t - g_t^1 - R_t^*\tilde u_t^2\right), \ \text{a.e. } t\in[\tau,T],$$

$$\mathbf{E}_{1,\tau}^{0,\tilde u^2}\,\tilde u_t^2 = G_{2,t}^{-1}\mathbf{E}_{1,\tau}^{0,\tilde u^2}\left(B_{2,t}^* y_t - L_{2,t}\tilde x_t - g_t^2 - R_t\tilde u_t^1\right), \ \text{a.e. } t\in[\tau,T],$$

which imply

$$\mathbf{E}_{1,\tau}^{0,\tilde u^2}\left(G_{1,t}\tilde u_t^1 + R_t^*\tilde u_t^2\right) = \mathbf{E}_{1,\tau}^{0,\tilde u^2}\left(B_{1,t}^* y_t - L_{1,t}\tilde x_t - g_t^1\right), \ \text{a.e. } t\in[\tau,T],$$

$$\mathbf{E}_{1,\tau}^{0,\tilde u^2}\left(R_1\tilde u_t^1 + G_{2,t}\tilde u_t^2\right) = \mathbf{E}_{1,\tau}^{0,\tilde u^2}\left(B_{2,t}^* y_t - L_{2,t}\tilde x_t - g_t^2\right), \ \text{a.e. } t\in[\tau,T].$$

Using the notation introduced above, we can write

$$G_t \mathbf{E}_{1,\tau}^{0,\tilde{u}^2} \tilde{u}_t = \mathbf{E}_{1,\tau}^{0,\tilde{u}^2} \left(B_t^* y_t - L_t \tilde{x}_t - g_t\right), \text{ a.e. } t \in [\tau, T],$$

which implies

$$\mathbf{E}_{1,\tau}^{0,\tilde{u}^2} \tilde{u}_t = G_t^{-1} \mathbf{E}_{1,\tau}^{0,\tilde{u}^2} \left(B_t^* y_t - L_t \tilde{x}_t - g_t\right), \text{ a.e. } t \in [\tau, T]. \tag{4.10}$$

Substituting (4.10) in (4.1) and (3.7), for all $0 < \tau \leq t \leq T$, we have

$$\mathbf{E}_{1,\tau}^{0,\tilde{u}^2} \tilde{x}_t = \mathbf{E}_{1,\tau}^{0,\tilde{u}^2} \bigg(\mathcal{U}_{t,\tau} \tilde{x}_\tau + \int_\tau^t \mathcal{U}_{t,s}\, dm_s + \int_\tau^t \mathcal{U}_{t,s} \big(B_s G_s^{-1} (B_s^* y_s - L_s \tilde{x}_s - g_s) + b_s \big) ds \bigg), \tag{4.11}$$

$$\mathbf{E}_{1,\tau}^{0,\tilde{u}^2} y_t = -\mathbf{E}_{1,\tau}^{0,\tilde{u}^2} \bigg(\mathcal{U}_{T,t}^* (Q_T \tilde{x}_T + q) + \int_t^T \mathcal{U}_{s,t}^* \big(F_s \tilde{x}_s + L_s^* G_s^{-1} (B_s^* y_s - L_s \tilde{x}_s - g_s) + f_s \big) ds \bigg). \tag{4.12}$$

Applying Lemma 3.7 to (4.11)–(4.12), we obtain

$$\mathbf{E}_{1,\tau}^{0,\tilde{u}^2} y_t = -\mathbf{E}_{1,\tau}^{0,\tilde{u}^2} (Q_t \tilde{x}_t + \alpha_t), \; 0 < \tau < t \leq T, \tag{4.13}$$

where $\mathcal{Y} = \mathcal{P}_{-BG^{-1}(B^*Q+L)}(\mathcal{U})$, Q is a solution of equation (3.5) and α is defined by (3.11). Finally, writing $t = \tau$ in (4.13) and using it in (4.9),

$$\begin{aligned} \tilde{u}_t^1 = -\, & \left(G_{1,t} - R_t^* G_{2,t}^{-1} R_t\right)^{-1} \mathbf{E}_{1,t}^{0,\tilde{u}^2} \Big(\big((B_{1,t}^* - R_t^* G_{2,t}^{-1} B_{2,t}^*) Q_t \\ & + L_{1,t} - R_t^* G_{2,t}^{-1} L_{2,t} \big) \tilde{x}_t + \big(B_{1,t}^* - R_t^* G_{2,t}^{-1} B_{2,t}^* \big) \alpha_t \\ & + g_t^1 - R_t^* G_{2,t}^{-1} g_t^2 \Big), \text{ a.e. } t \in \mathbf{T}. \end{aligned} \tag{4.14}$$

Thus, we can state the following theorem.

Theorem 4.2. *Under conditions* $(\mathbf{G_1})$–$(\mathbf{G_4})$, $(\mathbf{RE})$ *and* $(\mathbf{G_5'})$, *let the pair* $(\tilde{u}^1, \tilde{u}^2) \in U_{\mathrm{ad}}$ *be a saddle point in the game* (4.1)–(4.4). *Denote*

$$\tilde{x} = x^{\tilde{u}^1, \tilde{u}^2}, \; \mathbf{E}_{1,t}^{\tilde{u}^1, \tilde{u}^2} = \mathbf{E}\left(\cdot \,\middle|\, z_1^{\tilde{u}^1, \tilde{u}^2, t} \right).$$

Then $\tilde{u}^1$ *has the form*

$$\begin{aligned} \tilde{u}_t^1 = -\, & \left(G_{1,t} - R_t^* G_{2,t}^{-1} R_t\right)^{-1} \mathbf{E}_{1,t}^{\tilde{u}^1,\tilde{u}^2} \Big(\big((B_{1,t}^* - R_t^* G_{2,t}^{-1} B_{2,t}^*) Q_t \\ & + L_{1,t} - R_t^* G_{2,t}^{-1} L_{2,t} \big) \tilde{x}_t + \big(B_{1,t}^* - R_t^* G_{2,t}^{-1} B_{2,t}^* \big) \alpha_t \\ & + g_t^1 - R_t^* G_{2,t}^{-1} g_t^2 \Big), \text{ a.e. } t \in \mathbf{T}, \end{aligned} \tag{4.15}$$

where $\mathcal{Y} = \mathcal{P}_{-BG^{-1}(B^*Q+L)}(\mathcal{U})$, Q *is a solution of equation* (3.5) *and* α *is defined by* (3.11).

Proof. By (4.8), $\mathbf{E}_{1,t}^{0,\tilde{u}^2} = \mathbf{E}_{1,t}^{\tilde{u}^1,\tilde{u}^2}$, $0 < t \leq T$. Hence, (4.15) follows from (4.14). □

Theorem 4.2 can be interpreted in the following way: since the second player is completely informed about the observations of the first player (this follows from condition $(\mathbf{G}_5'))$, he (she) can compute the $\tilde{u}^1$ component of a saddle point. But, generally, the first player can not do the same about $\tilde{u}^2$ as he (she) has worse observations.

4.3. Case 2: The Players Have the Same Observations

A mathematical condition corresponding to this case is formulated as

$$(\mathbf{G}_5'')\ \sigma\left(z_1^{u^1,u^2,t}\right) = \sigma\left(z_2^{u^1,u^2,t}\right) \text{ for all } (u^1,u^2)\in U_{\mathrm{ad}} \text{ and for all } 0<t\le T.$$

In particular, condition $(\mathbf{G}_6'')$ holds if

$$(\mathbf{G}_5)\ Z_1 = Z_2 = Z,\ n^1 = n^2 = n \in M_2(\mathbf{T},Z),\ c^1 = c^2 = c \in L_2(\mathbf{T}\times\Omega,Z),\\ C_1 = C_2 = C \in B_\infty(\mathbf{T},\mathcal{L}(X,Z)).$$

Obviously, condition $(\mathbf{G}_5'')$ implies $\mathbf{E}_{1,t}^{\tilde{u}^1,\tilde{u}^2} = \mathbf{E}_{2,t}^{\tilde{u}^1,\tilde{u}^2}$.

Theorem 4.3. *Under conditions* $(\mathbf{G}_1)$*–*$(\mathbf{G}_4)$*,* $(\mathbf{RE})$ *and* $(\mathbf{G}_5'')$*, let the pair* $\tilde{u} = (\tilde{u}^1,\tilde{u}^2)\in U_{\mathrm{ad}}$ *be a saddle point in the game* (4.1)*–*(4.4)*. Denote*

$$\tilde{u} = \begin{bmatrix} \tilde{u}^1 \\ \tilde{u}^2 \end{bmatrix},\ \tilde{x} = x^{\tilde{u}^1,\tilde{u}^2},\ \tilde{\mathbf{E}}_t = \mathbf{E}_{1,t}^{\tilde{u}^1,\tilde{u}^2} = \mathbf{E}_{2,t}^{\tilde{u}^1,\tilde{u}^2}.$$

Then $\tilde{u}$ *has the form*

$$\tilde{u}_t = -G_t^{-1}(B_t^*Q_t + L_t)\tilde{\mathbf{E}}_t\,\tilde{x}_t - G_t^{-1}B_t^*\tilde{\mathbf{E}}_t\,\alpha_t - G_t^{-1}\tilde{\mathbf{E}}_t\,g_t,\ \text{a.e. } t\in\mathbf{T}, \tag{4.16}$$

where $\mathcal{Y} = \mathcal{P}_{-BG^{-1}(B^*Q+L)}(\mathcal{U})$*,* Q *is a solution of equation* (3.5) *and* α *is defined by* (3.11)*. Moreover, this saddle point is unique if equation* (3.5) *has a unique solution.*

Proof. By Theorem 4.2, we have (4.15). Under condition $(\mathbf{G}_5'')$, Theorem 4.2 can be applied to the second player too. Therefore, we also have

$$\begin{aligned}\tilde{u}_t^2 = -&\left(G_{2,t} - R_tG_{1,t}^{-1}R_t^*\right)^{-1}\mathbf{E}_{2,t}^{\tilde{u}^1,\tilde{u}^2}\big(\big((B_{2,t}^* - R_tG_{1,t}^{-1}B_{1,t}^*)Q_t\\ &+ L_{2,t} - R_tG_{1,t}^{-1}L_{1,t}\big)\tilde{x}_t + \big(B_{2,t}^* - R_tG_{1,t}^{-1}B_{1,t}^*\big)\alpha_t\\ &+ g_t^2 - R_tG_{1,t}^{-1}g_t^1\big),\ \text{a.e. } t\in\mathbf{T}.\end{aligned} \tag{4.17}$$

Combining (4.15) and (4.17) and using Proposition 4.1, we obtain (4.16). Obviously, (4.16) uniquely determines the optimal control (if it exists) if equation (3.5) has a unique solution. □

Note that Theorem 4.3 determines the form of the saddle point under its existence. In the following two sections we will discuss the existence of a saddle point in two games involving the heat and wave equations.

4.4. Application to the Heating-Cooling Game

Assume that a string of unit length is given. Two players have access to every point on the string at any time moment. For the time period T, the first player wants to make the temperature of the string close to a certain value, say, to zero and the other one wants to make it much away from zero, both of them with minimum effort. At both end points of the string a constant temperature equal to zero is kept. Obviously, the temperature fluctuation along the string between the two end points will be modeled as wide band noise. Both players observe the average temperature over the string up to the current time moment with the measurement noise that will be modeled as white noise. Mathematically, this "heating-cooling" game can be formulated as

$$\begin{aligned} \frac{\partial}{\partial t} x_{t,\alpha} &= \frac{\partial^2}{\partial \alpha^2} x_{t,\alpha} + u^1_{t,\alpha} + u^2_{t,\alpha} + \varphi_t,\ 0 < t \le T, \\ & x_{0,\alpha} = f_\alpha,\ x_{t,0} = x_{t,1} = 0,\ 0 \le \alpha \le 1,\ 0 \le t \le T, \end{aligned} \tag{4.18}$$

$$z_t = \int_0^t \int_0^1 x_{s,\alpha}\, d\alpha\, ds + v_t,\ 0 \le t \le T, \tag{4.19}$$

$$J(u^1, u^2) = \mathbf{E} \int_0^1 x^2_{T,\alpha}\, d\alpha + \mathbf{E} \int_0^T \int_0^1 \left(\left(u^1_{t,\alpha}\right)^2 - \left(u^2_{t,\alpha}\right)^2 \right) d\alpha\, dt, \tag{4.20}$$

where the function x expresses the temperature of the string, u^1 and u^2 are controls of the first and second players, respectively, and z is the common observation process of the players. Here v is a standard Wiener process and φ is a wide band noise process represented in the form (see Bashirov [9])

$$\varphi_t = \int_{\max(0,t-\varepsilon)}^t \phi_{\theta - t}\, dw_\theta,\ 0 \le t \le T, \tag{4.21}$$

with $0 < \varepsilon < T$, $\phi \in L_2(-\varepsilon, 0)$ and w being a standard Wiener process such that w and v are independent.

Let $X = L_2(0,1)$. The second order differential operator $d^2/d\alpha^2$ defined on

$$D\left(d^2/d\alpha^2\right) = \{h \in X : (d^2/d\alpha^2)h \in X,\ h_0 = h_1 = 0\}$$

generates the strongly continuous semigroup $\mathcal{U}$ defined by

$$[\mathcal{U}_t h]_\alpha = \sum_{n=1}^\infty 2e^{-n^2\pi^2 t} \sin(n\pi\alpha) \int_0^1 h_\theta \sin(n\pi\theta)\, d\theta,\ 0 \le \alpha \le 1,\ t \ge 0,\ h \in X,$$

and satisfying $\mathcal{U}_t = \mathcal{U}_t^*$, $t \ge 0$.

Since $Q_T = I$, $F_s = 0$, $L_s = 0$ and

$$B_s G_s^{-1} B_s^* = \begin{bmatrix} I & I \end{bmatrix} \begin{bmatrix} I & 0 \\ 0 & -I \end{bmatrix} \begin{bmatrix} I \\ I \end{bmatrix} = 0,$$

for the game (4.18)–(4.20), one can easily observe that

$$\mathcal{Y}_t = \mathcal{U}_t, \ t \geq 0,$$

and the Riccati equation (3.5) is trivial, expressing its solution explicitly in the form

$$Q_t = \mathcal{U}_{2(T-t)}, \ 0 \leq t \leq T.$$

Hence, by Theorem 4.3, if there exists a saddle point in this game, then it is unique and has the components

$$\tilde{u}^1_{t,\cdot} = -\mathcal{U}_{2(T-t)} \tilde{\mathbf{E}}_t \tilde{x}_{t,\cdot} - \tilde{\mathbf{E}}_t \int_t^T \mathcal{U}_{2T-t-s} D\varphi_s \, ds, \ \text{a.e. } t \in \mathbf{T}, \tag{4.22}$$

$$\tilde{u}^2_{t,\cdot} = \mathcal{U}_{2(T-t)} \tilde{\mathbf{E}}_t \tilde{x}_{t,\cdot} + \tilde{\mathbf{E}}_t \int_t^T \mathcal{U}_{2T-t-s} D\varphi_s \, ds, \ \text{a.e. } t \in \mathbf{T}, \tag{4.23}$$

where $D : (-\infty, \infty) \to L_2(0,1)$ is such that $[Dh]_\alpha = h$ for $-\infty < h < \infty$ and for $0 \leq \alpha \leq 1$. Using (4.22)–(4.23) in (4.18), we obtain

$$\tilde{x}_{t,\cdot} = \mathcal{U}_t f + \int_0^t \mathcal{U}_{t-s} D\varphi_s \, ds, \ 0 \leq t \leq T. \tag{4.24}$$

Therefore, (4.22)–(4.23) can be rewritten as

$$\tilde{u}^1_{t,\cdot} = -\mathcal{U}_{2T-t} f - \tilde{\mathbf{E}}_t \int_0^T \mathcal{U}_{2T-t-s} D\varphi_s \, ds, \ \text{a.e. } t \in \mathbf{T}, \tag{4.25}$$

$$\tilde{u}^2_{t,\cdot} = \mathcal{U}_{2T-t} f + \tilde{\mathbf{E}}_t \int_0^T \mathcal{U}_{2T-t-s} D\varphi_s \, ds, \ \text{a.e. } t \in \mathbf{T}. \tag{4.26}$$

In order to show the existence of the saddle point, we will verify it on the pair $(\tilde{u}^1, \tilde{u}^2)$. At first, note that from Bashirov [9], it follows that conditions $(\mathbf{E_1})$–$(\mathbf{E_2})$ for the game (4.18)–(4.20) hold. Hence, $(\tilde{u}^1, \tilde{u}^2) \in U_{\text{ad}}$. From Theorem 3.13, it is clear that

$$J(\tilde{u}^1, \tilde{u}^2) \leq J(u^1, \tilde{u}^2), \ (u^1, \tilde{u}^2) \in U_{\text{ad}}. \tag{4.27}$$

Let us show that

$$J(\tilde{u}^1, u^2) \leq J(\tilde{u}^1, \tilde{u}^2), \ (\tilde{u}^1, u^2) \in U_{\text{ad}}. \tag{4.28}$$

For, take any u^2 with $(\tilde{u}^1, \tilde{u}^2 + u^2) \in U_{\mathrm{ad}}$. Then

$$
\begin{aligned}
&J(\tilde{u}^1, \tilde{u}^2 + u^2) - J(\tilde{u}^1, \tilde{u}^2) \\
&\quad = 2\mathbf{E}\Big\langle \mathcal{U}_T f + \int_0^T \mathcal{U}_{T-s} D\varphi_s \, ds, \int_0^T \mathcal{U}_{T-t} u^2_{t,\cdot} \, dt \Big\rangle_X \\
&\qquad + \mathbf{E}\Big\| \int_0^T \mathcal{U}_{T-t} u^2_{t,\cdot} \, dt \Big\|_X^2 \\
&\qquad - 2\mathbf{E} \int_0^T \langle \tilde{u}^2_{t,\cdot}, u^2_{t,\cdot} \rangle_X \, dt \\
&\qquad - \mathbf{E} \int_0^T \|u^2_{t,\cdot}\|_X^2 \, dt \\
&\quad \leq 2\mathbf{E} \int_0^T \Big\langle \mathcal{U}_{2T-t} f + \int_0^T \mathcal{U}_{2T-t-s} D\varphi_s \, ds, u^2_{t,\cdot} \Big\rangle_X \, dt \\
&\qquad - 2\mathbf{E} \int_0^T \Big\langle \mathcal{U}_{2T-t} f + \tilde{\mathbf{E}}_t \int_0^T \mathcal{U}_{2T-t-s} D\varphi_s \, ds, u^2_{t,\cdot} \Big\rangle_X \, dt \\
&\qquad + \mathbf{E} \int_0^T \|u^2_{t,\cdot}\|_X^2 \, dt \Big(\int_0^T \|\mathcal{U}_{T-t}\|^2_{\mathcal{L}(X)} \, dt - 1 \Big) \\
&\quad = \mathbf{E} \int_0^T \|u^2_{t,\cdot}\|_X^2 \, dt \Big(\int_0^T \|\mathcal{U}_t\|^2_{\mathcal{L}(X)} \, dt - 1 \Big).
\end{aligned}
$$

Since

$$
\begin{aligned}
\int_0^T \|\mathcal{U}_t\|^2_{\mathcal{L}(X)} \, dt &= \int_0^T \sup_{\|h\|=1} \sum_{n=1}^{\infty} 2e^{-2n^2\pi^2 t} \Big(\int_0^1 h_\alpha \sin(n\pi\alpha) \, d\alpha \Big)^2 dt \\
&\leq \int_0^T e^{-2\pi^2 t} \, dt \\
&= \frac{1}{2\pi^2} \Big(1 - e^{-2\pi^2 T} \Big),
\end{aligned}
$$

we obtain

$$
\int_0^T \|\mathcal{U}_t\|^2_{\mathcal{L}(X)} \, dt - 1 \leq \frac{1}{2\pi^2} \Big(1 - 2\pi^2 - e^{-2\pi^2 T} \Big) \leq 0.
$$

Thus

$$
J(\tilde{u}^1, \tilde{u}^2 + u^2) - J(\tilde{u}^1, \tilde{u}^2) \leq 0
$$

proving that a saddle point in the game (4.18)–(4.20) exists.

Thus, the pair $(\tilde{u}^1, \tilde{u}^2)$ defined by (4.22)–(4.23) or (4.25)–(4.26) is a unique saddle point and $J(\tilde{u}^1, \tilde{u}^2) = \mathbf{E}\|\tilde{x}_{T,\cdot}\|_X^2$ where $\tilde{x}$ is defined by (4.24).

4.5. Application to a Game with a Vibrating String

Now consider a game involving a vibrating string of unit length: let the state system be given by

$$\frac{\partial^2}{\partial t^2}x_{t,\alpha} = \frac{\partial^2}{\partial \alpha^2}x_{t,\alpha} + u^1_{t,\alpha} + u^2_{t,\alpha} + \varphi_t,\ 0 < t \le 1,$$
$$x_{0,\alpha} = f_\alpha,\ \frac{\partial}{\partial t}x_{t,\alpha}\Big|_{t=0} = g_\alpha,$$
$$x_{t,0} = x_{t,1} = 0,\ 0 \le \alpha \le 1,\ 0 \le t \le 1, \tag{4.29}$$

the observation system by

$$z_t = \int_0^t \int_0^1 x_{s,\alpha}\, d\alpha\, ds + v_t,\ 0 \le t \le 1, \tag{4.30}$$

and the functional by

$$J(u^1,u^2) = \mathbf{E}\int_0^1 \left(\left(\frac{\partial}{\partial t}x_{t,\alpha}\Big|_{t=1}\right)^2 + \left(\frac{\partial}{\partial \alpha}x_{t,\alpha}\Big|_{t=1}\right)^2\right)d\alpha$$
$$+\ \mathbf{E}\int_0^1\int_0^1 \left(\left(u^1_{t,\alpha}\right)^2 - \left(u^2_{t,\alpha}\right)^2\right) d\alpha\, dt. \tag{4.31}$$

Here $x_{t,\alpha}$ expresses the displacement of the string at the point α and at the time moment t, u^1 is a control of the first player which tries to minimize the functional (4.31) and u^2 is a control of the second player maximizing (4.31). Both players observe the process z defined by (4.30). We assume that φ is a wide band noise process defined by (4.21) with $0 < \varepsilon < 1$, $\phi \in L_2(-\varepsilon, 0)$ with w being a standard Wiener process such that it is independent of the standard Wiener process v.

With each function $\xi \in L_2(0,1)$, we will associate the sequence $\{\breve{\xi}_n\}$ of Fourier coefficients in the half-range Fourier sine expansion

$$\xi_\alpha = \sum_{n=1}^{\infty} \breve{\xi}_n \sqrt{2}\sin(n\pi\alpha).$$

Let X be the Hilbert space of the functions

$$h = \begin{bmatrix} \xi \\ \eta \end{bmatrix} \in L_2(0,1) \times L_2(0,1)$$

with

$$\sum_{n=1}^{\infty} n^2 \breve{\xi}_n^2 < \infty$$

endowed with the scalar product

$$\langle h, h' \rangle_X = \left\langle \begin{bmatrix} \xi \\ \eta \end{bmatrix}, \begin{bmatrix} \xi' \\ \eta' \end{bmatrix} \right\rangle = \sum_{n=1}^{\infty} \left(n^2\pi^2 \breve{\xi}_n \breve{\xi}'_n + \breve{\eta}_n \breve{\eta}'_n\right).$$

Denote

$$A = \begin{bmatrix} 0 & I \\ d^2/d\alpha^2 & 0 \end{bmatrix}$$

where $d^2/d\alpha^2$ is the second order differential operator defined on

$$D(d^2/d\alpha^2) = \{\xi \in L_2(0,1) : (d^2/d\alpha^2)\xi \in L_2(0,1),\ \xi_0 = \xi_1 = 0\}$$

and I is the identity operator on $L_2(0,1)$. It is well known (see Curtain and Zwart [11] and Zabczyk [12]) that A generates the strongly continuous semigroup $\mathcal{U}_t$, $t \geq 0$. Moreover, $\mathcal{U}$ is a contraction semigroup, i.e., $\|\mathcal{U}_t\|_{\mathcal{L}(X)} \leq 1$ for all $t \geq 0$, and it satisfies $\mathcal{U}_t^* = \mathcal{U}_t^{-1}$, $t \geq 0$. We will extend $\mathcal{U}$ to the real line by letting $\mathcal{U}_{-t} = \mathcal{U}_t^{-1}$, $t > 0$.

Let

$$y_{t,\alpha} = \begin{bmatrix} x_{t,\alpha} \\ (\partial/\partial t)x_{t,\alpha} \end{bmatrix},\ y_{0,\cdot} = \begin{bmatrix} f \\ g \end{bmatrix} \in X.$$

Note that, for $\xi \in L_2(0,1)$ with $(d/d\alpha)\xi \in L_2(0,1)$, we have

$$\xi_\alpha = \sum_{n=1}^{\infty} \breve{\xi}_n \sqrt{2}\sin(n\pi\alpha) \Rightarrow \frac{d}{d\alpha}\xi_\alpha = \sum_{n=1}^{\infty} n\pi\breve{\xi}_n\sqrt{2}\cos(n\pi\alpha).$$

Hence, by Parseval's identity,

$$\int_0^1 \xi_\alpha^2\, d\alpha = \sum_{n=1}^{\infty} \breve{\xi}_n^2 \quad \text{and} \quad \int_0^1 \left(\frac{\partial}{\partial\alpha}\xi_\alpha\right)^2 d\alpha = \sum_{n=1}^{\infty} n^2\pi^2\breve{\xi}_n^2.$$

Therefore, the game (4.29)–(4.31) can be written as

$$y_{t,\cdot} = \mathcal{U}_t y_{0,\cdot} + \int_0^t \mathcal{U}_{t-s}(\tilde{I}u^1_{s,\cdot} + \tilde{I}u^2_{s,\cdot})\, ds + \int_0^t \mathcal{U}_{t-s}\tilde{I}D\varphi_s\, ds,\ 0 \leq t \leq 1, \tag{4.32}$$

$$z_t = \int_0^t \int_0^1 \begin{bmatrix} I & 0 \end{bmatrix} y_{s,\alpha}\, d\alpha\, ds + v_t,\ 0 \leq t \leq 1, \tag{4.33}$$

$$J(u^1,u^2) = \mathbf{E}\|y_{1,\cdot}\|_X^2 + \mathbf{E}\int_0^1 \left(\|u^1_{t,\cdot}\|^2_{L_2(0,1)} - \|u^2_{t,\cdot}\|^2_{L_2(0,1)}\right)dt, \tag{4.34}$$

where D is as in the example from Section 4.4 and

$$\tilde{I} = \begin{bmatrix} 0 \\ I \end{bmatrix} \in \mathcal{L}(L_2(0,1), X).$$

Since $Q_T = I$, $F_s = 0$, $L_s = 0$ and

$$B_s G_s^{-1} B_s^* = \begin{bmatrix} \tilde{I} & \tilde{I} \end{bmatrix} \begin{bmatrix} I & 0 \\ 0 & -I \end{bmatrix} \begin{bmatrix} \tilde{I}^* \\ \tilde{I}^* \end{bmatrix} = 0,$$

for the game (4.32)–(4.34), we obtain $\mathcal{Y}_t = \mathcal{U}_t$, $t \geq 0$, and $Q_t = I$, $0 \leq t \leq 1$. Hence, by Theorem 4.3, if there exists a saddle point in this game, then it is unique and has the components

$$\tilde{u}^1_{t,\cdot} = -\tilde{I}^*\tilde{\mathbf{E}}_t\tilde{y}_{t,\cdot} - \tilde{I}^*\tilde{\mathbf{E}}_t\int_t^1 \mathcal{U}_{t-s}\tilde{I}D\varphi_s\, ds,\ \text{a.e. } t \in [0,1], \tag{4.35}$$

$$\tilde{u}^2_{t,\cdot} = \tilde{I}^*\tilde{\mathbf{E}}_t\tilde{y}_{t,\cdot} + \tilde{I}^*\tilde{\mathbf{E}}_t\int_t^1 \mathcal{U}_{t-s}\tilde{I}D\varphi_s\, ds,\ \text{a.e. } t \in [0,1]. \tag{4.36}$$

Substituting (4.35)–(4.36) in (4.32), we obtain

$$\tilde{y}_{t,\cdot} = \mathcal{U}_t y_{0,\cdot} + \int_0^t \mathcal{U}_{t-s} \tilde{I} D \varphi_s \, ds, \ \ 0 \le t \le 1. \tag{4.37}$$

Therefore, (4.35)–(4.36) can be rewritten as

$$\tilde{u}^1_{t,\cdot} = -\tilde{I}^* \mathcal{U}_t y_{0,\cdot} - \tilde{\mathbf{E}}_t \int_0^1 \tilde{I}^* \mathcal{U}_{t-s} \tilde{I} D \varphi_s \, ds, \ \text{ a.e. } t \in [0,1], \tag{4.38}$$

$$\tilde{u}^2_{t,\cdot} = \tilde{I}^* \mathcal{U}_t y_{0,\cdot} + \tilde{\mathbf{E}}_t \int_0^1 \tilde{I}^* \mathcal{U}_{t-s} \tilde{I} D \varphi_s \, ds, \ \text{ a.e. } t \in [0,1]. \tag{4.39}$$

In order to show the existence of the saddle point, we will verify it on the pair $(\tilde{u}^1, \tilde{u}^2)$. From Bashirov [9], it follows that conditions $(\mathbf{E_1})$–$(\mathbf{E_2})$ for the game (4.32)–(4.34) hold. Hence, $(\tilde{u}^1, \tilde{u}^2) \in U_{\text{ad}}$. From Theorem 3.13, it is clear that the inequality (4.27) holds. Let us show the inequality (4.28) for the game (4.32)–(4.34). For, take any u^2 with $(\tilde{u}^1, \tilde{u}^2 + u^2) \in U_{\text{ad}}$. Then

$$\begin{aligned}
&J(\tilde{u}^1, \tilde{u}^2 + u^2) - J(\tilde{u}^1, \tilde{u}^2) \\
&\quad = 2\mathbf{E}\Big\langle \mathcal{U}_1 y_{0,\cdot} + \int_0^1 \mathcal{U}_{1-s} \tilde{I} D \varphi_s \, ds, \int_0^1 \mathcal{U}_{1-t} \tilde{I} u^2_{t,\cdot} \, dt \Big\rangle_X \\
&\quad\quad + \mathbf{E}\Big\| \int_0^1 \mathcal{U}_{1-t} \tilde{I} u^2_{t,\cdot} \, dt \Big\|_X^2 \\
&\quad\quad - 2\mathbf{E} \int_0^1 \langle \tilde{u}^2_{t,\cdot}, u^2_{t,\cdot} \rangle_{L_2(0,1)} \, dt - \mathbf{E} \int_0^1 \| u^2_{t,\cdot} \|^2_{L_2(0,1)} \, dt \\
&\quad \le 2\mathbf{E} \int_0^1 \Big\langle \tilde{I}^* \mathcal{U}_t y_{0,\cdot} + \int_0^1 \tilde{I}^* \mathcal{U}_{t-s} \tilde{I} D \varphi_s \, ds, u^2_{t,\cdot} \Big\rangle_{L_2(0,1)} dt \\
&\quad\quad - 2\mathbf{E} \int_0^1 \Big\langle \tilde{I}^* \mathcal{U}^*_t y_{0,\cdot} + \tilde{\mathbf{E}}_t \int_0^1 \tilde{I}^* \mathcal{U}^*_{s-t} \tilde{I} D \varphi_s \, ds, u^2_{t,\cdot} \Big\rangle_{L_2(0,1)} dt \\
&\quad\quad + \mathbf{E} \int_0^1 \| u^2_{t,\cdot} \|^2_{L_2(0,1)} \, dt \Big(\int_0^1 \| \mathcal{U}_{1-t} \tilde{I} \|^2_{\mathcal{L}(L_2(0,1),X)} \, dt - 1 \Big) \le 0.
\end{aligned}$$

Hence, the inequality (4.28) holds, proving that a saddle point in the game (4.32)–(4.34) or (4.29)–(4.31) exists.

Thus, the pair $(\tilde{u}^1, \tilde{u}^2)$ defined by (4.35)–(4.36) or (4.38)–(4.39) is a unique saddle point and $J(\tilde{u}^1, \tilde{u}^2) = \mathbf{E}\|\tilde{y}_{1,\cdot}\|_X^2$ where $\tilde{y}$ is defined by (4.37).

References

[1] L.S. Pontryagin, V.G. Boltyanskii, R.V. Gamkrelidze and E.F. Mischenko, *The Mathematical Theory of Optimal Processes*, Independence Publisher, New York, 1962.

[2] R. Bellman, *Dynamic Programming*, Princeton University Press, Princeton, 1957.

[3] W.M. Wonham, On the separation theorem of stochastic control, *SIAM J. Control*, **6** (1968), 312–326.

[4] A. Bensoussan and M. Viot, Optimal control of stochastic linear distributed parameter systems, *SIAM J. Control*, **13** (1975), 904–926.

[5] R.F. Curtain and A. Ichikawa, The separation principle for stochastic evolution equations, *SIAM J. Control and Optimization*, **15** (1977), 367–383.

[6] A.E. Bashirov, Optimal control of partially observable systems with arbitrarily dependent noises, *Stochastics*, **17** (1986), 163–205.

[7] E. Hille and R.S. Phillips, *Functional Analysis and Semigroups*, Amer. Math. Soc. Coll. Publ., Vol. 31, Providence, R.I., 1957.

[8] R.S. Bucy and P.D.Joseph, *Filtering for Stochastic Processes with Application to Guidance*, Interscience Publishers, New York, 1968.

[9] A.E. Bashirov, On linear filtering under dependent wide band noises, *Stochastics*, **23** (1988), 413–437.

[10] A.E. Bashirov and L.R. Mishne, On linear filtering under dependent white noises, *Stochastics and Stochastics Reports*, **35** (1991), 1–23.

[11] R.F. Curtain and H.J. Zwart, *An Introduction to Infinite Dimensional Linear Systems Theoty*, Springer Verlag,New York, 1995.

[12] J. Zabczyk, *Mathematical Control Theory: An Introduction*, Birkhäuser, Berlin, 1992.

Department of Mathematics
Eastern Mediterranean University
Gazimagusa
Mersin 10
Turkey
e-mail: agamirza.bashirov@emu.edu.tr

Progress in Probability, Vol. 53, 29–66

Stochastic Games with Risk Sensitive Payoffs for N Players

A. Bensoussan and J. Frehse

1. Introduction

In this paper we extend some of the results of the authors on stochastic differential games[1], to the case of risk sensitive payoffs. We consider that the stochastic process describing the system is stopped at the exit of a domain $\mathcal{O}$ of R^n. As in the non-stopped case (finite time horizon), the risk sensitive parameter cannot be arbitrary, see H. Nagai [10], A. Bensoussan, J. Frehse, H. Nagai [5] for the finite horizon case.

On the other hand, the fact that we consider risk sensitive payoffs prevents making use of discount factors in the cost functions. The system of Bellman equations related to the value functions has Dirichlet boundary conditions, and does not contain zero order terms. This complicates obtaining L^∞ bounds, since maximum principle arguments cannot be obtained easily. Other methods, already used by the authors to solve ergodic control problems, apply conveniently to the present case.

Our approach to solving the stochastic differential games, taken as usual in the sense of Nash, is to prove regularity results for the solution of the system of Bellman equations. Then a standard verification argument can be used.

2. Setting of the Problem

Let

$$\Omega = C^o([0,\infty); R^n), \quad \mathcal{A} = \text{ Borel } \sigma\text{-algebra on } \Omega. \tag{2.1}$$

The elements of Ω are denoted by $\omega = \omega(t)$ and we equip Ω with a probability law P such that

$$\omega(t) \text{is a standardized } n\text{-dimensional Wiener process.} \tag{2.2}$$

We then set

$$\mathcal{F}^t = \sigma\{\omega(s) \leq \omega \in \Omega\}. \tag{2.3}$$

[1]A. Bensoussan – J. Frehse [2], [3]

A trajectory starting at x is simply

$$x(t;\omega) = x + \omega(t). \tag{2.4}$$

We now consider N players, each of them acting through a control $v_\nu(t), \nu = 1, \dots, N$. We assume that

$$v(t) = (v_1(t), \dots, v_N(t)), \text{ an adapted process with bounded values in } R^{nN}, \tag{2.5}$$

which we call an admissible control.

Let also

$$g(x) \text{ be a measurable bounded function with values in } R^n. \tag{2.6}$$

To a pair $x, v(t)$, where $v(t)$ is a control vector as above, we associate the process

$$\beta_{x,v}(t) = g(x(t)) + \sum_\mu v_\mu(t) \tag{2.7}$$

and the probability $P_{x,v}$ such that

$$\frac{dP_{x,v}}{dP}|_{\mathcal{F}^t} = \exp\left\{\int_0^t \beta_{x,v}(s)dw(s) - \frac{1}{2}\int_0^t |\beta_{x,v}(s)|^2 ds\right\}. \tag{2.8}$$

From the Girsanov theorem, if we introduce the process

$$w_{x,v}(t) = \omega(t) - \int_0^t \beta_{x,v}(s)ds, \tag{2.9}$$

then the system $\Omega, \mathcal{A}, \mathcal{F}^t, P_{x,v}, w_{x,v}(t)$ forms a probability system in which $w_{x,v}(t)$ is an $\mathcal{F}^t$ standardized Wiener process. Note that from (2.9) one has

$$dx = (g(x(t)) + \sum_\mu v_\mu(t))dt + dw_{x,v}(t), \; x(0) = x. \tag{2.10}$$

Let now

$$\mathcal{O} = \text{an open smooth bounded domain of } R^n \tag{2.11}$$

and let

$$\tau_x = \inf\{t | x(t) \notin \mathcal{O}\}. \tag{2.12}$$

We shall stop the process $x(t)$ at the exit of the domain $\mathcal{O}$, and to save the notation, we shall still denote by $x(t)$ the stopped process. Let also

$$f_\nu(x) \text{ be a scalar measurable bounded function.} \tag{2.13}$$

In the sequel, we shall use the notation

$$v(t) = (v_\nu(t), v^\nu(t)), \tag{2.14}$$

where $v^\nu(t)$ represents all components different from v_ν. We also shall use the notation

$$\bar{v}_\nu(t) = \sum_{\mu \neq \nu} v_\mu(t). \tag{2.15}$$

The payoff of player ν is given by

$$\begin{aligned} J_\nu(x, v(\cdot)) &= J_\nu(x, v_\nu(\cdot), v^\nu(\cdot)) \\ &= \frac{1}{\delta} \log E_{x,v} \exp \delta \Big[\int_0^{T_x} (f_\nu(x(t)) + \frac{1}{2} |v_\nu(t)|^2 + \theta v_\nu(t) \cdot \bar{v}_\nu(t)) dt \Big], \end{aligned} \tag{2.16}$$

where δ stands for the risk parameter. Note that as δ tends to 0, this reduces to

$$E_{x,v} \Big[\int_0^{T_x} (f_\nu(x(t)) + \frac{1}{2} |v_\nu(t)|^2 + \theta v_\nu(t) \cdot \bar{v}_\nu(t)) dt \Big],$$

which is the payoff considered in [4] except for the fact that we do not have a discount factor anymore. If there is only one player, then the parameter θ is irrelevant.

A Nash point for the game defined by the functionals (2.16) is a control $\hat{v}(\cdot)$ such that

$$J_\nu(x, \hat{v}_\nu, \hat{v}^\nu) \leq J_\nu(x, v_\nu, \hat{v}^\nu), \ \ \forall \nu \tag{2.17}$$

for any admissible control $v(\cdot)$. See J. Nash [11], J.P. Aubin [1] for the concept of Nash point.

As indicated in the introduction, the method to prove the existence of a Nash point will be to consider a system of Bellman equations for the value function of the game. This means that, for a convenient control $\hat{v}$, possibly depending on x (the initial state), the functions

$$u_\nu(x) = J_\nu(x, \hat{v}) \tag{2.18}$$

are the solutions of a system of partial differential equations. This system will allow us to characterize optimal feedbacks for the N players. The proof of optimality will be performed by a verification argument. A key point is to obtain sufficient regularity properties for the value functions, otherwise, it is not possible to obtain feedback. Techniques of partial differential equations are instrumental in obtaining these necessary regularity properties.

3. Preliminaries

3.1. Lagrangians

We introduce the Lagrangians by the following definition:

$$L_\nu(v, p) = \frac{1}{2} |v_\nu|^2 + \theta v_\nu \cdot \hat{v}_\nu + p_\nu \cdot \sum_\mu v_\mu, \tag{3.1}$$

where

$$p = (p_1, \ldots, p_N) \in R^{nN},$$
$$v = (v_1, \ldots, v_N) \in R^{nN},$$

and, consistent with the notation (2.15), we denote

$$\bar{v}_\nu = \sum_{\mu \neq \nu} v_\mu, \tag{3.2}$$

$$\bar{p}_\nu = \sum_{\mu \neq \nu} p_\mu. \tag{3.3}$$

The first step is to consider, for a given p, a Nash point in ν for the function $L_\nu(v, p)$. Clearly, the following conditions must hold (by differentiation) for such a Nash point $\nu(p)$:

$$v_\nu(p) + \theta \bar{v}_\nu(p) + p_\nu = 0, \tag{3.4}$$

provided

$$\theta \neq 1,\ \theta \neq -\frac{1}{N-1'}. \tag{3.5}$$

It is easy to check that the system (3.4) has a unique solution given by the formulas

$$v_\nu(p) = \frac{\theta \sum_\mu p_\mu}{(1-\theta)(1+(N-1)\theta)} - \frac{p_\nu}{1-\theta}. \tag{3.6}$$

We note also the complementary formulas

$$\bar{v}_\nu(p) = \frac{-\sum_\mu p_\mu}{(1-\theta)(1+(N-1)\theta)} + \frac{p_\nu}{1-\theta}. \tag{3.7}$$

Then, we can define the quantities

$$L_\nu(p) = L_\nu(v(p), p). \tag{3.8}$$

It is useful to also express, from (3.6) and (3.7), the vectors p_ν in terms of $v(p)$ as

$$p_\nu = -v_\nu(p) - \theta \bar{v}_\nu(p), \tag{3.9}$$

and also

$$\bar{p}_\nu = -(N-1)\theta v_\nu(p) + (-N\theta + 2\theta - 1)\bar{v}_\nu(p). \tag{3.10}$$

In particular, we can write

$$L_\nu(p) = -\frac{1}{2}|v_\nu(p)|^2 + p_\nu \cdot \bar{v}_\nu(p). \tag{3.11}$$

Using (3.6), (3.7) in (3.11), after easy calculations we obtain

$$\begin{aligned} L_\nu(p) = &-\frac{\theta^2}{2(1-\theta)^2(1+(N-1)\theta)^2}\Big|\sum_\mu p_\mu\Big|^2 \\ &+\frac{1-2\theta}{2(1-\theta)^2}|p_\nu|^2 + \frac{2\theta-1}{(1-\theta)^2(1+(N-1)\theta)} p_\nu \cdot \sum_\mu p_\mu. \end{aligned} \tag{3.12}$$

3.2. More developments on Lagrangians

We continue some useful developments on Lagrangians. We first note by summing up (3.12) that

$$\begin{aligned}\sum_\nu L_\nu(p) &= \frac{\theta^2(3N-4)-2\theta(N-3)-2}{2(1-\theta)^2(1+(N-1)\theta)^2}\Big|\sum_\mu p_\mu\Big|^2 \\ &+\frac{1-2\theta}{2(1-\theta)^2}\sum_\nu |p_\nu|^2 \\ &= \frac{-2(N-1)^2\theta^3+(N^2-3N+1)\theta^2+2\theta-1}{2(1-\theta)^2(1+(N-1)\theta)^2}\sum_\nu |p_\nu|^2 \\ &+2\frac{(3N-4)\theta^2-2(N-3)\theta-2}{2(1-\theta)^2(1+(N-1)\theta)^2}\sum_{\mu,\nu_{\mu\neq\nu}} p_\mu\cdot p_\nu.\end{aligned} \tag{3.13}$$

We shall be interested in guaranteeing the property

$$\sum_\nu L_\nu(p) \geq c_0|p|^2, \quad \forall p, \quad c_0>0 \tag{3.14}$$

or

$$\sum_\nu L_\nu(p) \leq -c_1|p|^2, \quad \forall p, \quad c_1>0. \tag{3.15}$$

From considerations on quadratic forms, in order to obtain (3.14) it will be enough to have

$$\begin{aligned}F(\theta) &= -(N-1)^2\theta^3+(N^2-3N+1)\frac{\theta^2}{2}+\theta-\frac{1}{2} \\ &-(N-1)\Big|\Big(\frac{3N}{2}-2\Big)\theta^2-(N-3)\theta-1\Big|>0 \\ &\text{and} \quad c_0=\frac{F(\theta)}{(1-\theta)^2(1+(N-1)\theta)^2}.\end{aligned} \tag{3.16}$$

Similarly, to obtain (3.15) it will be sufficient to assert that

$$\begin{aligned}\hat{F}(\theta) &= -(N-1)^2\theta^3+(N^2-3N+1)\frac{\theta^2}{2}+\theta-\frac{1}{2} \\ &+(N-1)\Big|\Big(\frac{3N}{2}-2\Big)\theta^2-(N-3)\theta-1\Big|<0 \\ &\text{and} \quad c_1=\frac{-\hat{F}(\theta)}{(1-\theta)^2(1+(N-1)\theta)^2}.\end{aligned} \tag{3.17}$$

We shall characterize the values of θ for which (3.16) or (3.17) hold. Considering the two possibilities for the absolute value, we are led to introducing the following two functions (after reduction of terms).

$$F_1(\theta) = -(N-1)^2\theta^3+\frac{-2N^2+4N-3}{2}\theta^2+\theta(N^2-4N+4)+N-\frac{3}{2}, \tag{3.18}$$

$$F_2(\theta) = -(N-1)^2\theta^3+\frac{4N^2-10N+5}{2}\theta^2-\theta(N^2-4N+2)-N+\frac{1}{2}. \tag{3.19}$$

To make the values of $F(\theta)$ and $\hat{F}(\theta)$ explicit we consider the two roots of

$$(3N-4)\frac{\theta^2}{2} - (N-3)\theta - 1,$$

namely

$$\theta_0 = \frac{N-3-\sqrt{N^2+1}}{3N-4}, \theta_0 > -1 (\text{ for } N \geq 3),\ \ \theta_0 < -1 \text{ (for } N=2), \tag{3.20}$$

$$\theta_0' = \frac{N-3+\sqrt{N^2+1}}{3N-4}, \quad \theta_0' < 1, \tag{3.21}$$

and we have

$$\begin{aligned} F(\theta) &= F_1(\theta), \quad \text{if } \theta \leq \theta_0 \ \text{ or } \theta \geq \theta_0', \\ F(\theta) &= F_2(\theta), \quad \text{if } \quad \theta_0 \leq \theta \leq \theta_0', \end{aligned} \tag{3.22}$$

$$\begin{aligned} \hat{F}(\theta) &= F_1(\theta), \quad \text{if } \theta_0 \leq \theta \leq \theta_0', \\ \hat{F}(\theta) &= F_2(\theta), \quad \text{if } \theta \leq \theta_0 \text{ or } \theta \geq \theta_0'. \end{aligned} \tag{3.23}$$

Fortunately, $F_2(\theta)$ is quite simple. Indeed, from (3.19), it is easy to check that

$$F_2(\theta) = (\theta-1)^2(-(N-1)^2\theta - N + \frac{1}{2}). \tag{3.24}$$

Let then

$$\bar{\theta} = -\frac{N-\frac{1}{2}}{(N-1)^2} \tag{3.25}$$

and note that

$$\theta_0 < \bar{\theta} < 0 \quad (-1 < \theta_0 < \bar{\theta}, \text{ for } N \geq 3), (\theta_0 < \bar{\theta} < -1, \text{ for } N=2). \tag{3.26}$$

Clearly

$$\begin{aligned} F_2(\theta) > 0 \quad &\text{if} \quad \theta \leq \bar{\theta}, \\ F_2(\theta) \leq 0 \quad &\text{if} \quad \theta \geq \bar{\theta}. \end{aligned} \tag{3.27}$$

From the definition (3.22) of $F(\theta)$, we can immediately assert that

$$\begin{aligned} F(\theta) > 0 \quad &\text{for} \quad \theta \leq \theta < \bar{\theta}, \\ F(\theta) \leq 0 \quad &\text{for} \quad \bar{\theta} \leq \theta \leq \theta_0', \end{aligned} \tag{3.28}$$

and similarly from (3.23)

$$\begin{aligned} \hat{F}(\theta) < 0 \quad &\text{for} \quad \theta \geq \theta_0', \quad \theta \neq 1, \\ \hat{F}(\theta) > 0 \quad &\text{for} \quad \theta \leq \theta_0'. \end{aligned} \tag{3.29}$$

To proceed, we must study the sign of $F_1(\theta)$, which is less simple.

However the case $N=2$ is very simple and particular, since in this case one has

$$F_1(\theta) = (\theta+1)^2(-\theta + \frac{1}{2}). \tag{3.30}$$

Note that for $N=2$, we have

$$\theta_0 = \frac{-1-\sqrt{5}}{2}, \quad \theta_0' = \frac{-1+\sqrt{5}}{2}, \quad \bar{\theta} = -\frac{3}{2}. \tag{3.31}$$

Thus we can easily complete (3.28)–(3.29) for $N = 2$:

$$\begin{aligned} F(\theta) = F_1(\theta) > 0 \quad \text{for} \quad \theta \leq \theta_0, \\ F(\theta) = F_1(\theta) < 0 \quad \text{for} \quad \theta \geq \theta_0', \end{aligned} \tag{3.32}$$

$$\begin{aligned} \hat{F}(\theta) = F_1(\theta) < 0 \quad \text{for} \quad \tfrac{1}{2} < \theta \leq \theta_0', \\ \hat{F}(\theta) = F_1(\theta) \geq 0 \quad \text{for} \quad \theta_0 \leq \theta \leq \tfrac{1}{2}. \end{aligned} \tag{3.33}$$

Therefore we can state

$$\text{for } N = 2, \begin{cases} F(\theta) > 0 & \text{is equivalent to} \quad \theta < -\frac{3}{2} = \bar{\theta} \\ \hat{F}(\theta) < 0 & \text{is equivalent to} \quad \theta > \frac{1}{2} = \theta'', \theta \neq 1. \end{cases} \tag{3.34}$$

To study $F_1(\theta)$, for $N \geq 3$, we consider

$$F_1'(\theta) = -3(N-1)^2\theta^2 + (-2N^2 + 4N - 3)\theta + N^2 - 4N + 4,$$

whose roots are θ_1, θ_1' given by the formulas

$$\theta_1 = \frac{-2N^2 + 4N - 3 - \sqrt{16N^4 - 88N^3 + 184N^2 - 168N + 57}}{6(N-1)^2}, \tag{3.35}$$

$$\theta_1' = \frac{-2N^2 + 4N - 3 + \sqrt{16N^4 - 88N^3 + 184N^2 - 168N + 57}}{6(N-1)^2}. \tag{3.36}$$

We have the configuration

$$-1 < \theta_1 < \theta_0 < \bar{\theta} < 0 < \theta_1' < \theta_0' < 1 \tag{3.37}$$

and $F_1'(\theta) < 0$ for $\theta < \theta_1$, and $\theta > \theta_1'$, whereas $F_1'(\theta) \geq 0$ for $\theta_1 \leq \theta \leq \theta_1'$. Note also that

$$\begin{aligned} F_1(-\infty) = +\infty, \ F_1(-1) \leq 0, \ F_1(\theta_0) = F_2(\theta_0) > 0, \\ F_1(\theta_0') = F_2(\theta_0') < 0, \ F_1(+\infty) = -\infty. \end{aligned} \tag{3.38}$$

From the sign of $F_1(\theta)$, it also follows that

$$\begin{aligned} F_1(\theta_1) < 0, \quad \text{minimum of} \quad F_1(\theta), \\ F_1(\theta_1') > 0, \quad \text{maximum of} \quad F_1(\theta). \end{aligned} \tag{3.39}$$

Therefore $F_1(\theta)$ has three roots, two being negative, θ', θ'' and one positive θ''', with the location

$$\theta' \leq -1 \quad , \theta_1 < \theta'' < \theta_0, \theta_1' < \frac{1}{2} < \theta''' < \theta_0', \tag{3.40}$$

and we can conclude easily that

$$F(\theta) > 0 \quad \text{for} \quad \theta \leq \theta' \quad \text{or} \quad \theta'' < \theta < \bar{\theta}, \tag{3.41}$$

$$\hat{F}(\theta) \leq 0 \quad \text{for} \quad \theta > \theta''' \quad \theta \neq 1. \tag{3.42}$$

We then state the

Lemma 3.1. *For $N = 2$, (3.14) holds whenever $\theta < -\frac{3}{2}$, and (3.15) holds whenever $\theta > \frac{1}{2}, \theta \neq 1$. For $N \geq 3$, considering the numbers $\theta', \theta'', \theta''', \bar{\theta}$, where θ', θ'' are the two negative roots of $F_1(\theta), \theta'''$ the positive root and $\bar{\theta}$ is given by (3.25), the property (3.14) holds when $\theta < \theta'$ or $\theta'' < \theta < \bar{\theta}$, and the property (3.15) holds when $\theta > \theta''', \theta \neq 1$.*

Remark 3.2. In (3.5) we had excluded the values $\theta = 1$, and $\theta = -\frac{1}{N-1}$. Since $\bar{\theta} < -\frac{1}{N-1} < 0$ the value $-\frac{1}{N-1}$ is out of the validity intervals defined in Lemma 3.1. The value $\theta = 1$, valid for (3.15), but not for (3.14), has to be excluded.

Remark 3.3. For $N = 3$, we have

$$\theta' = -1, \theta'' = \frac{-1 - \sqrt{97}}{16}, \quad \theta''' = \frac{-1 + \sqrt{97}}{16}, \quad \theta_1 = \frac{-9 + \sqrt{129}}{24},$$

$$\theta_0 = -\frac{\sqrt{10}}{5}, \quad \theta_0' = -\frac{\sqrt{10}}{5}, \quad \bar{\theta} = -\frac{5}{8}.$$

3.3. Other properties

From formula (3.12), we can deduce, by using Young's inequality,

$$L_\nu(p) \leq \frac{1 - 2\theta}{2\theta^2} |p_\nu|^2. \tag{3.43}$$

We proceed now with a different estimate. Note first that from (3.7) we can write

$$\bar{v} = \bar{v}_\nu(p) = \frac{(N-1)\theta p_\nu - \bar{p}\nu}{(1-\theta)(1 + (N-1)\theta)}. \tag{3.44}$$

By analogy with the formula (3.44), we consider a similar combination of the Lagrangians, namely

$$\tilde{L}_\nu(p) = (N-1)\theta L_\nu(p) - \sum_{\mu \neq \nu} L_\mu(p). \tag{3.45}$$

Consider (3.9) and (3.11) which yields

$$\begin{aligned} L_\nu(p) &= -\frac{1}{2}|v_\nu|^2 - v_\nu \bar{v}_\nu - \theta |\bar{v}_\nu|^2 \\ &-\frac{1}{2}|v_\nu + \bar{v}_\nu|^2 + (\frac{1}{2} - \theta)|\bar{v}_\nu|^2, \end{aligned} \tag{3.46}$$

hence as easily seen:

$$\begin{aligned}\tilde{L}_\nu(p) &= (N-1)\theta\Big(\frac{1}{2}-\theta\Big)|v_\nu|^2 + \frac{1}{2}(N-1)(1-\theta)|v_\nu+\bar{v}_\nu|^2 \\ &\quad -\Big(\frac{1}{2}-\theta\Big)\sum_{\mu\neq\nu}|\bar{v}_\mu|^2 \\ &= (N-1)\theta\Big(\frac{1}{2}-\theta\Big)|\bar{v}_\nu|^2 + \frac{1-\theta}{2(N-1)}\Big|\sum_\mu \bar{v}_\mu\Big|^2 \\ &\quad -\Big(\frac{1}{2}-\theta\Big)\sum_{\mu\neq\nu}|\bar{v}_\mu|^2 \\ &= \Big[(N-1)\theta\Big(\frac{1}{2}-\theta\Big)+\frac{1}{2}\frac{1-\theta}{N-1}\Big]|\bar{v}_\nu|^2 + \frac{1}{2}\frac{1-\theta}{N-1}\Big|\sum_{\mu\neq\nu}\bar{v}_\mu\Big|^2 \\ &\quad +\frac{1-\theta}{N-1}\bar{v}_\nu\sum_{\mu\neq\nu}\bar{v}_\mu + \Big(\theta-\frac{1}{2}\Big)\sum_{\mu\neq\nu}|\bar{v}_\mu|^2.\end{aligned} \tag{3.47}$$

Using

$$\Big|\sum_{\mu\neq\nu}\bar{v}_\mu\Big|^2 \leq (N-1)\sum_{\mu\neq\nu}|\bar{v}_\mu|^2 \tag{3.48}$$

and assuming

$$\theta \geq \frac{1}{2}, \tag{3.49}$$

we deduce

$$\begin{aligned}\tilde{L}_\nu(p) &\geq \Big[(N-1)\theta\Big(\frac{1}{2}-\theta\Big)+\frac{1}{2}\frac{1-\theta}{N-1}\Big]|\bar{v}_\nu|^2 + \frac{\theta}{2(N-1)}\Big|\sum_{\mu\neq\nu}\bar{v}_\mu\Big|^2 \\ &\quad +\frac{1-\theta}{N-1}\bar{v}_\nu\sum_{\mu\neq\nu}\bar{v}_\mu \\ &\geq -\frac{2\theta-1}{2\theta(N-1)}(\theta^2(N-1)^2+\theta-1)|\bar{v}_\nu|^2.\end{aligned} \tag{3.50}$$

So we state the

Lemma 3.4. *When $\theta \geq \frac{1}{2}$, one has the property*

$$\tilde{L}_\nu(p) \geq -\frac{1}{2}|\bar{v}_\nu|^2. \tag{3.51}$$

□

In the sequel we shall use (3.43) and (3.14) together when $\theta < 0$ (θ satisfying the conditions of Lemma (3.1)) and (3.15), (3.51) together when $\theta > 0$ (in fact $\theta > \theta''' > \frac{1}{2}, \theta \neq 1$).

3.4. Taking account of the risk factor

The risk factor will imply a perturbation of the Lagrangian, namely $L_\nu(p)$ has to be replaced by

$$M_\nu(p) = L_\nu(p) + \frac{\delta}{2}|p_\nu|^2. \tag{3.52}$$

The property (3.43) is clearly unchanged:

$$M_\nu(p) \le \Big(\frac{1-2\theta}{2\theta^2} + \frac{\delta}{2}\Big)|p_\nu|^2. \tag{3.53}$$

The property (3.14) is improved, since

$$\sum_\nu M_\nu(p) = \sum_\nu L_\nu(p) + \frac{\delta}{2}|p|^2, \tag{3.54}$$

and if (3.14) holds, a fortiori

$$\sum_\nu M_\nu(p) \ge c_0|p|^2. \tag{3.55}$$

Since the risk factor here helps, it modifies the discussion on θ. From formula (3.13), what matters now is to have

$$\begin{aligned} F_\delta(\theta) = \frac{\delta}{2}(1-\theta)^2(1+(N-1)\theta)^2 - (N-1)^2\theta^3 + (N^2-3N+1)\frac{\theta^2}{2} + \theta - \frac{1}{2} \\ -(N-1)|(\frac{3N}{2}-2)\theta^2 - (N-3)\theta - 1| > 0. \end{aligned} \tag{3.56}$$

Define

$$F_{1\delta}(\theta) = F_1(\theta) + \frac{\delta}{2}(1-\theta)^2(1+(N-1)\theta)^2, \tag{3.57}$$

$$F_{2\delta}(\theta) = F_2(\theta) + \frac{\delta}{2}(1-\theta)^2(1+(N-1)\theta)^2. \tag{3.58}$$

We have

$$\begin{aligned} F_\delta(\theta) &= F_{1\delta}(\theta) \quad \text{if} \quad \theta \le \theta_0 \text{ or } \theta \ge \theta_0', \\ F_\delta(\theta) &= F_{2\delta}(\theta) \quad \text{if} \quad \theta_0 \le \theta \le \theta_0'. \end{aligned} \tag{3.59}$$

The function $F_{2\delta}(\theta)$ is easy to compute (using (3.24)):

$$F_{2\delta}(\theta) = \frac{(\theta-1)^2}{2}[\delta(N-1)^2\theta^2 + 2(N-1)\theta(\delta-(N-1)) - 2N + 1 + \delta], \tag{3.60}$$

hence the two roots not equal to 1 are

$$\bar\theta_\delta = \frac{1 - \frac{\delta}{N-1} - \sqrt{1 + \frac{\delta}{(N-1)^2}}}{\delta}, \qquad \bar\theta_\delta' = \frac{1 - \frac{\delta}{N-1} + \sqrt{1 + \frac{\delta}{(N-1)^2}}}{\delta}, \tag{3.61}$$

and $\bar\theta < \bar\theta_\delta$. Therefore

$$\begin{aligned} F_{2\delta}(\theta) > 0 \quad &\text{if} \quad \theta < \bar\theta_\delta \text{ or } \theta > \bar\theta'_\delta, \theta \ne 1, \\ F_{2\delta}(\theta) \le 0 \quad &\text{if} \quad \bar\theta_\delta \le \theta \le \bar\theta'_\delta. \end{aligned} \tag{3.62}$$

We need next to study $F_{1\delta}(\theta)$. We assume $N \ge 3$. Note that

$$F_{1\delta}'(\theta) = F_1'(\theta) + \delta(\theta-1)(1+(N-1)\theta)(2\theta(N-1)-(N-2)). \tag{3.63}$$

For $\theta < -\dfrac{1}{N-1}$, we have $F'_{1\delta}(\theta) < F'_1(\theta)$. Hence

$$F'_{1\delta}(\theta) < 0 \quad \text{for} \quad \theta \leq \theta_1.$$

Noting that

$$\theta'_1 < \frac{N-2}{2(N-1)} < \theta'_0,$$

we have also

$$F'_{1\delta}(\theta) < 0 \quad \text{for} \quad \frac{N-2}{2(N-1)} \leq \theta \leq 1.$$

Similarly, since $\bar{\theta} < -\dfrac{1}{N-1}$, one has also

$$F'_{1\delta}(\theta) > 0 \quad \text{for} \quad -\frac{1}{N-1} \leq \theta \leq \theta'_1.$$

Furthermore $F'_{1\delta}(\theta) > 0$ for θ sufficiently large > 1. Therefore $F'_{1\delta}(\theta)$ has necessarily three roots, which we denote $\theta_{1\delta}, \theta'_{1\delta}, \theta''_{1\delta}$. We can also assert that

$$\begin{aligned} &\theta_1 < \theta_{1\delta} < -\frac{1}{N-1}, \\ &\theta'_1 < \theta'_{1\delta} < \frac{N-2}{2(N-1)} \quad , \quad \theta''_{1\delta} > 1. \end{aligned} \tag{3.64}$$

Moreover as $\delta \to 0$, $\theta_{1\delta} \to \theta_1$, $\theta'_{1\delta} \to \theta'_1$, $\theta''_{1\delta} \to +\infty$ and as $\delta \to \infty$, $\theta_{1\delta} \to -\dfrac{1}{N-1}$, $\theta'_{1\delta} \to \frac{N-2}{2(N-1)}$, $\theta''_{1\delta} \to 1$.

Hence $F_{1\delta}(\theta)$ has two local minima, $\theta_{1\delta}$ and $\theta''_{1\delta}$, and one local maximum $\theta'_{1\delta}$. Note the following properties:

$$\begin{aligned} &F_{1\delta}(\theta) > 0, \quad \theta \leq \theta' \quad \text{and} \quad \theta'' \leq \theta \leq \theta''', \\ &F_{1\delta}(1) = F_1(1) = -N^2 + N < 0, \\ &F_{1\delta}(-\infty) = F_{1\delta} = (+\infty) = +\infty. \end{aligned} \tag{3.65}$$

Therefore the value at the local minimum $\theta_{1\delta}$ is strictly negative, and thus $F_{1\delta}$ has two positive roots $\theta'''_{1\delta}, \theta'''_{2\delta}$. Necessarily, since $F_{1\delta}(\theta''') > 0$, we have

$$\theta''' < \theta'''_{1\delta} < 1 < \theta'''_{2\delta}. \tag{3.66}$$

The value at the local minimum $\theta_{1\delta}$ is not necessarily negative. It is so if δ is sufficiently small, since $\theta_{1\delta}$ is close to θ_1, and $F_{1\delta}(\theta_{1\delta}) \sim F_1(\theta_1) < 0$. In that case there will be two negative roots of $F_{1\delta}(\theta)$, denoted by $\theta'_\delta, \theta''_\delta$, and from (3.65) necessarily one has

$$\theta' < \theta'_\delta < \theta_{1\delta} < \theta''_\delta < \theta''. \tag{3.67}$$

Lemma 3.5. *If θ belongs to the validity intervals defined in* (3.70)*, then we have* (3.53), (3.55).

On the other hand if $\theta_0' < \bar{\theta}_\delta'$ and $\theta \in (\theta_{1\delta}''', \bar{\theta}_\delta')$ and satisfies the properties (3.78), (3.80)*, then* (3.71), (3.72) *are satisfied.*

4. Nonlinear Systems of Partial Differential Equations

4.1. Setting of the problem

Here we consider the system of equations

$$\begin{aligned} -\frac{1}{2}\Delta u_\nu &= H_\nu(x, Du), \\ u_\nu|_{\partial\mathcal{O}} &= 0. \end{aligned} \tag{4.1}$$

The functions $H_\nu(x,p)$ satisfy first

$$H_\nu(x,p) \text{ are Caratheodory functions.} \tag{4.2}$$

We shall make an important use of linear manipulations on equations (4.1). Consider an $N \times N$ matrix Γ which is invertible, and define

$$H^\Gamma(x,p) = \Gamma H(x, \Gamma^{-1}p), \tag{4.3}$$

where $H(x,p)$ represents the vector $H_\nu(x,p)$. Setting

$$z = \Gamma u, \tag{4.4}$$

we see that z is the solution of

$$\begin{aligned} -\tfrac{1}{2}\Delta z_\nu &= H_\nu^\Gamma(x, Dz), \\ z_\nu|_{\partial\mathcal{O}} &= 0. \end{aligned} \tag{4.5}$$

So our original problem (4.1) is imbedded in a family of equivalent problems, indexed by the transformation Γ. All matrices Γ to be considered will be *invertible*, so we shall not mention it explicitly.

A matrix Γ satisfies the maximum principle if

$$\Gamma u \geq 0 \Rightarrow u \geq 0. \tag{4.6}$$

We begin by stating two alternative sets of assumptions:

$$\sum_\nu H_\nu(x,p) \geq -\lambda. \tag{4.7}$$

There exists a matrix Γ which satisfies the maximum principle and $H_\nu^\Gamma(x,p) \leq \lambda_\nu + \lambda_\nu^0|p_\nu|^2, \lambda_\nu\lambda_\nu^0$ not too large. (4.8)

or

$$\sum_\nu H_\nu(x,p) \leq \lambda. \tag{4.9}$$

There exists a matrix Γ which satisfies the maximum principle and $H_\nu^\Gamma(x,p) \geq -\lambda_\nu - \lambda_\nu^0|p_\nu|^2, \lambda_\nu\lambda_\nu^0$ not too large. (4.10)

We furthermore assume

$$\begin{aligned}&\text{There exists a matrix } \Gamma \text{ such that}\\ &H_\nu^\Gamma(x,p) = Q(x,p)\cdot p_\nu + H_\nu^0(x,p)\end{aligned} \tag{4.11}$$

with

$$|Q(x,p)| \leq k + K|p| \tag{4.12}$$

$$|H_\nu^0(x,p)| \leq k_\nu + K_\nu \sum_{\mu\leq\nu} |p_\mu|^2. \tag{4.13}$$

Remark 4.1. If we pick $\nu = N$ in (4.13) we obtain

$$|H_N^0(x,p)| \leq k_N + K_N|p|^2,$$

which is a general quadratic growth assumption. So if $H(x,p)$ has a general quadratic growth, it is sufficient to check (4.11), (4.12), (4.13) for $\nu = 1\ldots N-1$. We may define

$$H_N^0(x,p) = H_N^\Gamma(x,p) - Q(x,p)p_N, \tag{4.14}$$

and (4.11), (4.13) will be satisfied automatically. □

Our objective is to prove the

Theorem 4.2. *We assume that the functions $H_\nu(x,p)$ satisfy* (4.2), (4.11), (4.12), (4.13) *and one or the other of the two sets of assumptions* (4.7), (4.8) *or* (4.9), (4.10). *Then there exists a solution of* (4.1) *which is in $W^{2,s}(\mathcal{O})$ for all s such that $2 \leq s < \infty$.*

The proof will be done by explaining how to obtain a priori estimates first, then an approximation argument will be used. We shall consider the Green function associated with any point $\xi \in \mathcal{O}$, corresponding to the operator $-\frac{1}{2}\Delta$, called G^ξ. It is the solution of the equation (written formally)

$$\begin{aligned}-\frac{1}{2}\Delta G^\xi &= \delta(x-\xi),\\ G^\xi|_{\partial\mathcal{O}} &= 0.\end{aligned} \tag{4.15}$$

The Green function is positive and satisfies the estimates

$$\|G^\xi\|_{L^q(\mathcal{O})} \leq C, \quad \forall\xi, \quad 1 \leq q < \frac{n}{n-2}, \tag{4.16}$$

$$\|G^\xi\|_{W^{1,r}(\mathcal{O})} \leq C, \quad \forall\xi, \quad 1 \leq r < \frac{n}{n-1}. \tag{4.17}$$

We shall denote by $\|G\|_{L^q}, \|G\|_{W^{1,r}}$ the bounds on the right-hand side of (4.16), (4.17).

Since Γ satisfies the maximum principle, this implies

$$u_\nu \geq -C, \tag{4.30}$$

which together with (4.27) means again

$$\|u_\nu\|_\infty \leq C. \tag{4.31}$$

4.3. H_0^1 estimates

To obtain the H_0^1 estimate, we shall make use of the special structure (4.11), (4.12), (4.13). We omit the notation Γ, to simplify the writing, and by virtue of the L^∞ estimate, we have

$$|u_\nu(x)| \leq \rho. \tag{4.32}$$

To obtain an a priori estimate for H_0^1, one uses a specific test function. Set

$$\beta(s) = e^s - s - 1$$

and

$$F = \prod_{\nu=1}^{N} \exp \beta(\gamma_\nu u_\nu),$$

where γ_ν is a positive constant to be defined later. We have

$$DF = F \sum_{\nu=1}^{N} \gamma_\nu \beta'(\gamma_\nu u_\nu) D u_\nu.$$

We test (4.1) with $F\gamma_\nu\beta'(\gamma_\nu u_\nu)$, which vanishes on the boundary, integrate by parts and add up. We get

$$\sum_\nu \frac{1}{2} \int_{\mathcal{O}} \gamma_\nu^2 |Du_\nu|^2 e^{\gamma_\nu u_\nu} F dx + \frac{1}{2} \int_{\mathcal{O}} \frac{|DF|^2}{F} dx$$
$$= \int_{\mathcal{O}} Q \cdot DF dx + \int_{\mathcal{O}} \sum_\nu \gamma_\nu H_\nu^0(Du) F (e^{\gamma_\nu u_\nu} - 1) dx,$$

hence also

$$\begin{aligned} \sum_\nu \frac{1}{2} \int_{\mathcal{O}} \gamma_\nu^2 |Du_\nu|^2 e^{\gamma_\nu u_\nu} F dx &\leq \frac{1}{2} \int_{\mathcal{O}} F Q \cdot Q dx \\ + \int_{\mathcal{O}} \sum_\nu \gamma_\nu H_\nu^0(Du) F (e^{\gamma_\nu u_\nu} &- 1) dx. \end{aligned} \tag{4.33}$$

In order to get comparable terms on both sides, we introduce the function

$$X = \prod_{\nu=1}^{N} (\exp \beta(\gamma_\nu u_\nu) + \exp \beta(-\gamma_\nu u_\nu)) \tag{4.34}$$

and the related quantities

$$X_\nu = X \frac{e^{\gamma_\nu u_\nu} \exp \beta(\gamma_\nu u_\nu) + e^{-\gamma_\nu u_\nu} \exp \beta(-\gamma_\nu u_\nu)}{\exp \beta(\gamma_\nu u_\nu) + \exp \beta(-\gamma_\nu u_\nu)},$$

$$\tilde{X}_\nu = X \frac{(e^{\gamma_\nu u_\nu} - 1) \exp \beta(\gamma_\nu u_\nu) - (e^{-\gamma_\nu u_\nu} - 1) \exp \beta(-\gamma_\nu u_\nu)}{\exp \beta(\gamma_\nu u_\nu) + \exp \beta(-\gamma_\nu u_\nu)}.$$

We have the inequalities

$$2^N \le X \le X_\nu \le X e^{\gamma_\nu |u_\nu|},$$
$$|\tilde{X}_\nu| \le X_\nu.$$

Applying the relations (4.33) with γ_ν changed at some time into $-\gamma_\nu$, and summing up the 2^N relations obtained in this way, we get the inequality

$$\frac{1}{2}\sum_\nu \int_{\mathcal{O}} \gamma_\nu^2 |Du_\nu|^2 X_\nu dx \le \frac{1}{2}\int_{\mathcal{O}} XQ\cdot Q dx$$
$$+\int_{\mathcal{O}} \sum_\nu \gamma_\nu H_\nu^0(Du)\tilde{X}_\nu dx,$$

hence

$$\frac{1}{2}\sum_\nu \int_{\mathcal{O}} \gamma_\nu^2 |Du_\nu|^2 X_\nu dx \le \frac{1}{2}\int_{\mathcal{O}} XQ\cdot Q dx$$
$$+\int_{\mathcal{O}} \sum_\nu \gamma_\nu |H_\nu^0(Du)| X_\nu dx$$
$$\le \frac{1}{2}\int_{\mathcal{O}} XQ\cdot Q dx + \int_{\mathcal{O}} \sum_\nu \gamma_\nu (k_\nu + K_\nu |Du_\nu|^2) X_\nu dx$$
$$+\int_{\mathcal{O}} \sum_\nu |Du_\nu|^2 \sum_{\mu>\nu} \gamma_\mu K_\mu X_\mu dx$$
$$\le \int_{\mathcal{O}} k^2 X dx + K^2 \sum_\nu \int_{\mathcal{O}} X_\nu |Du_\nu|^2 dx + \sum_\nu \gamma_\nu \int_{\mathcal{O}} (k_\nu + K_\nu |Du_\nu|^2) X_\nu dx$$
$$+\int_{\mathcal{O}} \sum_\nu |Du_\nu|^2 \sum_{\mu>\nu} \gamma_\mu K_\mu X_\mu dx.$$

We obtain

$$\sum_\nu \int_{\mathcal{O}} |Du_\nu|^2 X_\nu [\frac{1}{2}\gamma_\nu^2 - K^2 - \gamma_\nu K_\nu] \le \int_{\mathcal{O}} (k^2 X + \sum_\nu \gamma_\nu k_\nu X_\nu) dx$$
$$+\sum_\nu \int_{\mathcal{O}} |Du_\nu|^2 \sum_{\mu>\nu} \gamma_\mu K_\mu X_\mu dx.$$

Finally

$$\sum_\nu \int_{\mathcal{O}} X |Du_\nu|^2 [\frac{1}{2}\gamma_\nu^2 - K^2 - \gamma_\nu K_\nu - \sum_{\mu>\nu} \gamma_\mu k_\mu e^{\rho^\gamma \mu}] dx$$
$$\le \int_{\mathcal{O}} X(k^2 + \sum_\nu \gamma_\nu K_\nu e^{\rho^\gamma \nu}) dx.$$

Therefore if we choose the constants γ_ν so that

$$\frac{1}{2}\gamma_\nu^2 - K^2 - \gamma_\nu K_\nu - \sum_{\mu>\nu} \gamma_\mu K_\mu e^{\rho^\gamma \mu} > 0, \tag{4.35}$$

we get

$$\int_{\mathcal{O}} |Du|^2 dx \le K_0(\rho). \tag{4.36}$$

It is possible to choose the constants γ_ν in order to fulfill (4.35). This can be done backwards, starting with γ_N.

4.4. C^δ and $W^{1,p}$ estimates

The special structure permits us to obtain additional estimates in C^δ and $W^{1,p}, p > 2$. We perform first a calculation similar to that leading to the H_0^1 estimate. To u_ν we associate a constant c_ν which is arbitrary except

$$|c_\nu| \le \rho, \tag{4.37}$$

and set now

$$F = \prod_{\nu=1}^{N} \exp \beta(\gamma_\nu(u_\nu - c_\nu)).$$

Let also ψ be a function such that

$$\psi \ge 0, \psi \in C^1(\mathcal{O}), \psi|_{\partial\mathcal{O}} = 0 \quad \text{if and only if one of the constants } c_\nu \ne 0. \tag{4.38}$$

We test (4.1) with $F\gamma_\nu\beta'(\gamma_\nu(u_\nu - c_\nu))\psi$, which vanishes on the boundary of $\mathcal{O}$. We obtain instead of (4.33)

$$\begin{aligned}
&\sum_\nu \frac{1}{2}\int_{\mathcal{O}} \gamma_\nu^2 |Du_\nu|^2 e^{\gamma_\nu(u_\nu - c_\nu)} F\psi dx + \frac{1}{2}\int_{\mathcal{O}} DF \cdot D\psi dx \\
&\le \frac{1}{2}\int_{\mathcal{O}} FQ \cdot Q\psi dx + \int_{\mathcal{O}} \sum_\nu \gamma_\nu H_\nu^0(Du) F (e^{\gamma_\nu(u_\nu - c_\nu)} - 1)\psi dx.
\end{aligned} \tag{4.39}$$

Introduce now

$$X = \prod_{\nu=1}^{N} (\exp \beta(\gamma_\nu(u_\nu - c_\nu)) + \exp \beta(-\gamma_\nu(u_\nu - c_\nu)). \tag{4.40}$$

$$X_\nu = X \frac{e^{\gamma_\nu(u_\nu - c_\nu)} \exp \beta(\gamma_\nu(u_\nu - c_\nu)) + e^{-\gamma_\nu(u_\nu - c_\nu))} \exp \beta(-\gamma_\nu(u_\nu, c_\nu))}{\exp \beta(\gamma_\nu(u_\nu - c_\nu)) + \exp \beta(-\gamma_\nu(u_\nu - c_\nu))},$$

$$\tilde{X}_\nu = X \frac{(e^{\gamma_\nu(u_\nu - c_\nu)} - 1) \exp \beta(\gamma_\nu(u_\nu - c_\nu)) - (e^{-\gamma_\nu(u_\nu - c_\nu)} - 1) \exp \beta(-\gamma_\nu(u_\nu - c_\nu))}{\exp \beta(\gamma_\nu(u_\nu - c_\nu)) + \exp \beta(-\gamma_\nu(u_\nu - c_\nu))},$$

and note that

$$DX = \sum_\nu \gamma_\nu \tilde{X}_\nu D u_\nu. \tag{4.41}$$

So writing (4.39) for all combinations of γ_ν and $-\gamma_\nu$, and adding up, yields

$$\begin{aligned}
&\frac{1}{2}\sum_\nu \int_{\mathcal{O}} \gamma_\nu^2 |Du_\nu|^2 X\psi dx + \frac{1}{2}\int_{\mathcal{O}} DX \cdot D\psi dx \\
&\le \frac{1}{2}\int_{\mathcal{O}} XQ \cdot Q\psi dx + \sum_\nu \int_{\mathcal{O}} \gamma_\nu H_\nu^0(Du) \tilde{X}_\nu \psi dx.
\end{aligned} \tag{4.42}$$

Picking the constants γ_ν so that (4.35) holds, we deduce that there exist two positive constants k_0, K_0 so that

$$k_0 \int_{\mathcal{O}} |Du|^2 \psi dx + \int_{\mathcal{O}} DX \cdot D\psi dx \leq K_0 \int_{\mathcal{O}} \psi dx. \tag{4.43}$$

We first apply (4.43) as follows. Let $B_R(x_0), x_0 \in R^n$, be the ball of center x_0 and radius R. Let τ be a cut-off function, such that

$$\begin{cases} \tau = 1 & \text{on } B_1(0), \\ 0 & \text{outside } B_2(0), \end{cases}$$

and $0 \leq \tau \leq 1, \tau \in C^\infty$. We denote

$$\tau_R(x) = \tau(\frac{x - x_0}{R}).$$

We define the constants c_ν as follows:

$$C_\nu = C_\nu^R = \begin{cases} \dfrac{1}{|2R|} \displaystyle\int_{B_{2R}} u_\nu dx & \text{if } B_{2R} \subset \mathcal{O}, \\ 0 & \text{if } B_{2R} \cap (R^n - \mathcal{O}) \neq \emptyset, \end{cases} \tag{4.44}$$

and we take

$$\psi = \tau_R^2, \tag{4.45}$$

so (4.38) is verified. We extend u by 0 outside $\mathcal{O}$ and we note that

$$|\tilde{X}_\nu| \leq C|u_\nu - c_\nu^R|,$$

$$|DX| \leq C|u - c^R|\,|Du|,$$

where c^R represents the vector c_ν^R. Therefore we deduce from (4.43)

$$\int_{B_R} |Du|^2 dx \leq C \int_{B_{2R}} |Du| \frac{|u - c^R|}{R} dx + CR^n. \tag{4.46}$$

Such a property implies that

$$u_\nu \in W_0^{1,p}(\mathcal{O}), \quad 2 \leq p < 2 + \varepsilon. \tag{4.47}$$

It is a consequence of Gehring's result. Indeed, forming Hölder's inequality and Poincaré's inequality, we have

$$\int_{B_{2R}} |Du| \frac{|u - c^R|}{R} dx \leq \frac{C}{R} \left(\int_{B_{2R}} |Du|^{\frac{2n}{n+1}} dx \right)^{\frac{n+1}{2n}} \left(\int_{B_{2R}} |u - c^R|^{\frac{2n}{n-1}} dx \right)^{\frac{n-1}{2n}}$$

$$\leq \frac{C}{R} \left(\int_{B_{6R}} |Du|^{\frac{2n}{n+1}} dx \right)^{\frac{n+1}{2n}}.$$

Setting $z = |Du|^{\frac{2n}{n+1}}$, we thus have the inequality

$$\oint_{B_R} z^{\frac{n+1}{n}} dx \leq C \left(\oint_{B_{6R}} z dx \right)^{\frac{n+1}{n}} + C, \tag{4.48}$$

where

$$\oint_{B_R} = \frac{1}{R^n}\int_{B_R}.$$

This is the reverse Hölder's inequality, which implies Gehring's result, namely $z^{\frac{n+1}{n}+\varepsilon}$ is integrable for some positive ε, hence (4.47), see [7]. □

Obtaining C^δ is more delicate. The idea is to estimate the Morrey norm, that is to check that

$$\int_{B_R(x_0)} |Du|^2 dx \le KR^{n-2+2\delta}, \forall R, \forall x_0. \tag{4.49}$$

This implies C^δ. Indeed one relies on Morrey's result:

$$\sup_{\substack{x,y\\ x\neq y}} \frac{|u(x)-u(y)|}{|x-y|^\delta} \le C\left(\frac{\sup_{x_0,R}\int_{B_R(x_0)} |Du|^2 dx}{R^{n-2+2\delta}}\right)^{\frac{1}{2}}. \tag{4.50}$$

To check (4.49) we shall prove the inequality

$$\int_{B_R} |Du|^2|x-x_0|^{2-n}dx \le C\int_{B_{\sigma R}-B_R} |Du|^2|x-x_0|^{2-n}dx + CR^\beta, \tag{4.51}$$

where $\sigma > 2, \beta > 0$. One then relies on the Hole filling technique of Widman [12], to assert that (4.49) holds for some δ such that $2\delta < \beta$. Indeed (4.51) implies

$$\int_{B_R} |Du|^2|x-x_0|^{2-n}dx \le \theta\int_{B_{\sigma R}} |Du|^2|x-x_0|^{2-n}dx + CR^\beta \tag{4.52}$$

with $\theta < 1$. Setting

$$\varphi(R) = \frac{\int_{B_R} |Du|^2|x-x_0|^{2-n}dx}{R^{2\delta}}$$

with $2\delta < \beta, \quad \mu = \theta\sigma^{2\delta} < 1$, we deduce

$$\varphi(R) \le \mu\varphi(\sigma R) + C \quad , \text{ for } R < R_0,$$

which implies $\varphi(R) \le C$, provided $\varphi(R_0) < \infty$.

Proof of (4.51):

We consider the Green function $G = G^{x_0}$, a solution of

$$-\frac{1}{2}\Delta G^{x_0} = \delta(x-x_0) \qquad x_0 \in Q,$$

$$G|_{\partial Q} = 0,$$

where $Q \supset \bar{\mathcal{O}}$. We know that (see (4.16), (4.17))

$$G^{x_0} \in L^q(Q)\cap W_0^{1,r}(Q), \quad 1 \le q < \frac{n}{n-2}, 1 \le r < \frac{n}{n-1} \tag{4.53}$$

and also

$$c_0|x-x_0|^{2-n} \le G^{x_0} \le c_1|x-x_0|^{2-n} \tag{4.54}$$

in a neighborhood of x_0 strictly included in Q. So this estimate will be valid on $\bar{\mathcal{O}}$.

We apply (4.43) with

$$\psi = G^{x_0}\tau_R^2 \tag{4.55}$$

and

$$c_\nu = c_\nu^R = \begin{cases} \dfrac{1}{|B_{2R} - B_{\frac{R}{2}}|}\displaystyle\int_{B_{2R}-B_{\frac{R}{2}}} u_\nu dx & \text{if } B_{2R} \subset \mathcal{O}, \\ 0 & \text{if } B_{2R} \cap (R^n - \mathcal{O}) \neq \emptyset. \end{cases} \tag{4.56}$$

Consider the various terms in (4.43). We first have

$$k_0 \int_{\mathcal{O}} |Du|^2 G^{x_0}\tau_R^2 dx \geq C \int_{B_R} |Du|^2 |x - x_0|^{2-n} dx. \tag{4.57}$$

We next have

$$\int_{\mathcal{O}} G^{x_0}\tau_R^2 dx \leq C\, R^{\frac{n}{q'}} = CR^\beta, \beta < 2. \tag{4.58}$$

Next

$$D\psi = DG\tau_R^2 + 2G\tau_R D\tau_R.$$

Consider

$$I = \int_{\mathcal{O}} DXD\tau_R G\tau_R dx.$$

Using

$$|DX| \leq C|u - c^R|||Du| \tag{4.59}$$

we have

$$|I| \leq \int_{B_{2R}-B_R} |Du|\frac{|u - c^R|}{R} G dx$$

$$\leq C\int_{B_{2R}-B_R} |Du|^2|x - x_0|^{2-n}dx + C\int_{(B_{2R}-B_R)\cap\mathcal{O}} \frac{|u - c^R|^2}{R^2}|x - x_0|^{2-n}dx.$$

But

$$\int_{(B_{2R}-B_R)\cap\mathcal{O}} \frac{|u - c^R|^2}{R^2}|x - x_0|^{2-n}dx \leq \frac{C}{R^n}\int_{(B_{2R}-B_R)\cap\mathcal{O}} |u - c^R|^2 dx$$

$$\leq \frac{C}{R^n}\int_{(B_{2R}-B_{\frac{R}{2}})\cap\mathcal{O}} |u - c^R|^2 dx,$$

and by the Poincaré inequality

$$\int_{(B_{2R}-B_{\frac{R}{2}})\cap\mathcal{O}} |u - c^R|^2 \leq C\, R^2 \int_{(B_{\sigma R}-B_{\frac{R}{2}})\cap\mathcal{O}} |Du|^2 dx, \quad \sigma > 2.$$

Hence, collecting results, one has

$$|I| \leq C\int_{B_{\sigma R}-B_{\frac{R}{2}}} |Du|^2|x - x_0|^{2-n}dx. \tag{4.60}$$

The other term

$$II = \int_{\mathcal{O}} DX\ DG\tau_R^2 dx, \tag{4.61}$$

which involves DG, is more complicated to estimate. By changing X into $X - 2^N$ we have

$$II = \int_{\mathcal{O}} DG \cdot D((X - 2^N)\tau_R^2)dx - 2\int_{\mathcal{O}} DGD\tau_R(X - 2^N)\tau_R \cdot dx.$$

Without loss of generality we can assume that $B_{2R} \subset Q$, hence using $X \geq 2^N$, we have from the definition of the Green function

$$II > -C\int_{(B_{2R}-B_R)\cap\mathcal{O}} |DG|\frac{|u - c^R|^2}{R}\tau_R dx,$$

where we have used the property

$$|X^N - 2^N| \leq C|u - c^R|^2. \tag{4.62}$$

We get a lower estimate for II as follows:

$$II > -C\int_{(B_{2R}-B_R)\cap\mathcal{O}} \frac{|u - c^R|^2}{R^2}dx - C\int_{(B_{2R}-B_R)\cap\mathcal{O}} G^{-1}|DG|^2|u - c^R|^2\tau_R^2 dx.$$

The first term in the right-hand side is estimated by the right-hand side of (4.60). We thus need to estimate the term

$$III = \int_{(B_{2R}-B_R)\cap\mathcal{O}} G^{-1}|DG|^2|u - c^R|^2\tau_R^2 dx.$$

To estimate this quantity, we introduce a new cut-off function satisfying

$$\begin{cases} \xi = 0 & \text{for} \quad |x| \leq \frac{1}{2}, \\ \tau & \text{for} \quad |x| > 1, \end{cases}$$

and we set

$$\xi_R(x) = \xi(\frac{x - x_0}{R}).$$

Thus

$$\xi_R = \tau_R \quad \text{for} \quad B_{2R} - B_R.$$

From the Green function equation we deduce, by testing with

$$G^{-\frac{1}{2}}|u - c^R|^2\xi_R^2,$$

which vanishes in x_0, and on the boundary of $\mathcal{O}$,

$$\frac{1}{4}\int_{\mathcal{O}} |DG|^2G^{-\frac{3}{2}}|u - C^R|^2\xi_R^2 dx = \int_{\mathcal{O}} D(|u - c^R|^2\xi_R^2) \cdot DG\ G^{-\frac{1}{2}}dx. \tag{4.63}$$

On the other hand,

$$\int_{\mathcal{O}}(-\frac{1}{2}\Delta u_\nu)(u_\nu - c_\nu^R)G^{\frac{1}{2}}\xi_R^2 dx = \frac{1}{2}\int_{\mathcal{O}}|Du_\nu|^2 G^{\frac{1}{2}}\xi_R^2 dx$$

$$+\frac{1}{8}\int_{\mathcal{O}} D(\xi_R^2|u - c^R|^2)DG\, G^{\frac{1}{2}}dx$$

$$+\int_{\mathcal{O}} Du_\nu D\xi_R\xi_R(u_\nu - c_\nu^R)G^{\frac{1}{2}}dx - \frac{1}{4}\int_{\mathcal{O}} D\xi_R DG|u - c^R|^2 G^{-\frac{1}{2}}\xi_R dx,$$

so

$$\frac{1}{8}\int_{\mathcal{O}} D(\xi_R^2|u - c^R|^2)DG\, G^{-\frac{1}{2}}dx \le \int_{\mathcal{O}} H_\nu(u_\nu - c_\nu^R)G^{\frac{1}{2}}\xi_R^2 dx$$

$$-\int_{\mathcal{O}} Du_\nu D\xi_R\xi_R(u_\nu - c_\nu^R)G^{\frac{1}{2}}dx + \frac{1}{4}\int_{\mathcal{O}} D\xi_R DG|u - c^R|^2 G^{-\frac{1}{2}}\xi_R dx,$$

and from the quadratic growth of H_ν

$$\le CR^{n(1-\frac{1}{2q})} + CR^{\frac{2-n}{2}}\int_{B_{2R}-B_{\frac{R}{2}}}|Du|^2 dx + CR^{\frac{2n}{2}}\int_{(B_{2R}-B_{\frac{R}{2}})\cap\mathcal{O}}\frac{|u - c^R|^2}{R^2}dx$$

$$+\frac{1}{4}\int_{\mathcal{O}} D\xi_R DG|u - c^R|^2 G^{-\frac{1}{2}}\xi_R dx.$$

Furthermore,

$$\int_{\mathcal{O}} D\xi_R DG|u - c^R|^2 G^{-\frac{1}{2}}\xi_R dx \le C\delta\int_{\mathcal{O}}|DG|^2 G^{-\frac{3}{2}}|u - c^R|^2\xi_R^2 dx$$

$$+\frac{C}{\delta}R^{\frac{2-n}{2}}\int_{(B_{2R}-B_{\frac{R}{2}})\cap\mathcal{O}}\frac{|u - c^R|^2}{R^2}dx.$$

Collecting results and choosing δ sufficiently small, we have

$$\int_{\mathcal{O}}|DG|^2 G^{-\frac{3}{2}}|u - c^R|^2\xi_R^2 dx \le CR^{n(1-\frac{1}{2q})} + CR^{\frac{2-n}{2}}\int_{(B_{2R}-B_{\frac{R}{2}})\cap\mathcal{O}}\frac{|u - c^R|^2}{R^2}dx$$

$$+CR^{\frac{2-n}{2}}\int_{B_{2R}-B_{\frac{R}{2}}}|Du|^2 dx.$$

From Poincaré's inequality it follows that

$$\int_{\mathcal{O}}|DG|^2 G^{-\frac{3}{2}}|u - C^R|^2\xi_R^2 dx \le CR^{n(1-\frac{1}{2q})} + CR^{\frac{2-n}{2}}\int_{B_{\sigma R}-B_{\frac{R}{2}}}|Du|^2 dx.$$

Going back to the definition of III and recalling that $\xi_R = \tau_R$ on $B_{2R} - B_R$ we get

$$III \le CR^{\frac{2-n}{2}}\int_{\mathcal{O}} G^{-\frac{3}{2}}|DG|^2|u - c^R|^2\xi_R^2 dx,$$

and from the previous estimate

$$III \leq CR^{2-n}\int_{B_{\sigma R}-B_{\frac{R}{2}}}|Du|^2dx + CR^{1+\frac{n}{2q'}}$$

$$\leq C\int_{B_{\sigma R}-B_{\frac{R}{2}}}|Du|^2|x-x_0|^{2-n}dx + CR^{1+\frac{n}{2q'}}.$$

Therefore we can state the inequality

$$II \geq -\int_{B_{\sigma R}-B_{\frac{R}{2}}}|Du|^2|x-x_0|^{2-n}dx - CR^{1+\frac{n}{2q'}},$$

and from (4.43) we immediately get (4.51), with

$$\beta = \frac{n}{q}, \text{ since } \frac{n}{q} < 1 + \frac{n}{2q'}.$$ □

To apply the Hole filling technique, it remains to verify

$$\int_{\mathcal{O}}|Du|^2|x-x_0|^{2-n}dx < C. \tag{4.64}$$

For that we use again (4.43) with $c_\nu = 0$ and $\psi = G$. Since

$$\int_{\mathcal{O}} DX \cdot DGdx > 0,$$

the result (4.64) is obvious.

4.5. $W^{2,s}$ estimates

From the linear theory and $W^{1,p}, C^\delta$ estimates, for some $p > 2, \delta > 0$ we can derive $W^{2,s}$ estimates for any s. This is thanks to an interpolation result and a bootstrap argument. Indeed if $u_\nu \in W^{1,p_0} \cap C^\delta, \quad p_0 > 2, \delta > 0$, then $\Delta u_\nu \in L^{\frac{p_0}{2}}$ therefore $u_\nu \in W^{2,\frac{p_0}{2}}$ from linear regularity theory.

It follows from the Miranda–Nirenberg interpolation theorem (see [9]), that

$$u_\nu \subset W^{1,p_1} \quad \text{with} \quad \frac{1}{p_1} = \frac{1}{p_0} - \frac{\delta}{2n}$$

provided $p_0 < \dfrac{2n}{\delta}$, and thus $p_1 > p_0$. After a finite number of steps, we get $p_i \geq \frac{2n}{\delta}$, and it follows that

$$u_\nu \in W_0^{1,p}(\mathcal{O}), \quad p > 2n,$$

and from the linear theory again

$$u_\nu \in W_0^{2,s}(\mathcal{O}), \quad s > n.$$

From the Sobolev imbedding theorem, $u_\nu \in W^{1,r}, \forall r$ and from the linear theory again $u_\nu \in W^{2,s}, \forall s$.

4.6. The case of a positive zero order term

As a preliminary to problem (4.1) we consider the problem

$$\begin{aligned} -\tfrac{1}{2}\Delta u_\nu + \alpha u_\nu &= H_\nu(x, Du), \\ u_\nu|_{\partial\mathcal{O}} &= 0 \end{aligned} \tag{4.65}$$

with $\alpha > 0$. For such a problem, we can weaken the assumptions (4.7), (4.8) and (4.9), (4.10) as follows:

$$\sum_\nu H_\nu(x,p) \geq -\lambda - \bar{\lambda}|\sum_\nu p_\nu|^2, \quad \bar{\lambda} \geq 0, \lambda \geq 0. \tag{4.66}$$

There exists a matrix Γ which satisfies the maximum principle and $H_\nu^\Gamma(x,p) \leq \lambda_\nu + \lambda_\nu^0|p_\nu|^2, \quad \lambda_\nu, \lambda_\nu^0 \geq 0.$ (4.67)

$$\sum_\nu H_\nu(x,p) \leq \lambda + \bar{\lambda}|\sum_\nu p_\nu|^2, \quad \bar{\lambda} \geq 0, \lambda \geq 0. \tag{4.68}$$

There exists a matrix Γ which satisfies the maximum principle and $H^\Gamma(x,p) \geq -\lambda_\nu - \lambda_\nu^0|p_\nu|^2, \quad \lambda_\nu, \lambda_\nu^0 \geq 0.$ (4.69)

Theorem 4.3. *We assume that the functions $H_\nu(x,p)$ satisfy* (4.2), (4.11), (4.12), (4.13) *and one or the other of the two sets of assumptions* (4.66), (4.67) *or* (4.68), (4.69). *Then for $\alpha > 0$, there exists a solution of* (4.65) *which is in $W^{2,s}(\mathcal{O})$, for all s such that $2 \leq s < \infty$.* □

If one considers the developments of sections 4.2, 4.3, 4.4, 4.5, the only thing which fails is the treatment of the L^∞ estimate. But to recover the L^∞ estimate in the present framework is very easy since we can rely on Maximum Principle arguments. Indeed assuming first (4.66), (4.67), then we have

$$-\frac{1}{2}\Delta\sum_\nu u_\nu + \alpha(\sum_\nu u_\nu) \geq -\lambda - \bar{\lambda}|D\sum_\nu u_\nu|^2,$$

and if ξ is a point of negative minimum of $\sum_\nu u_\nu$, necessarily

$$D\sum_\nu u_\nu(\xi) = 0, \Delta\sum_\nu u_\nu(\xi) \geq 0,$$

hence

$$\sum_\nu u_\nu(\xi) \geq -\frac{\lambda}{\alpha}.$$

Therefore

$$\sum_\nu u_\nu(x) \geq -\frac{\lambda}{\alpha}. \tag{4.70}$$

Next considering

$$\tilde{u} = \Gamma u, \tag{4.71}$$

we get the system

$$-\frac{1}{2}\Delta\tilde{u}_\nu + \alpha\tilde{u}_\nu = H^\Gamma(x, D\tilde{u}),$$

hence from the assumption (4.67) it follows that

$$-\frac{1}{2}\Delta\tilde{u}_\nu + \alpha\tilde{u}_\nu \leq \lambda_\nu + \lambda_\nu^0|D\tilde{u}_\nu|^2,$$

and if ξ_ν is a point of positive maximum, necessarily

$$\tilde{u}_\nu(\xi_\nu) \leq \frac{\lambda_\nu}{\alpha}.$$

Therefore

$$\tilde{u}_\nu(x) \leq \frac{\lambda_\nu}{\alpha}, \tag{4.72}$$

and since Γ satisfies the maximum principle, we get

$$u_\nu(x) \leq \frac{1}{\alpha}\sum_\nu (\Gamma^{-1})_{\nu\mu}\lambda_\mu, \tag{4.73}$$

which combined with (4.70) yields the result.

The case of the assumption (4.68), (4.69) is treated in a similar way.

Once we know L^∞ estimates on αu_ν, then this term can be incorporated in the Lagrangian and the developments of sections 4.3, 4.4, 4.5 can be made. Of course, all the estimates on the functions u_ν depend on α.

Proof of Theorem 4.2:

One considers an approximation as follows:

$$\begin{aligned} -\frac{1}{2}\Delta u_\nu + \alpha u_\nu &= \frac{H_\nu(x, Du)}{1 + \varepsilon|H(x, Du)|} = H_\nu^\varepsilon(x \cdot Du), \\ u_\nu|\partial\mathcal{O} &= 0, \end{aligned} \tag{4.74}$$

where $H(x,p)$ represents the vector $H_\nu(x,p)$. Clearly the right-hand side is bounded in L^∞, by a constant depending on ε. To show that (4.74) has a solution in $W^{2,s}(\mathcal{O}) \cap W_0^{1,s}(\mathcal{O}), \forall 2 \leq s < \infty$, one relies on Schauder's fixed point theorem. Indeed consider the set, for s arbitrary,

$$K_\varepsilon = \{z \in (W_0^{1,s}(\mathcal{O}))^N \mid \|z\|_{(W^{2,s})^N} \leq C\}, \tag{4.75}$$

which is compact in $(W_0^{1,s}(\mathcal{O}))^N$. Consider next the map from $(W_0^{1,s})^N$ into itself, $T^\varepsilon\zeta = z$, by solving

$$\begin{aligned} -\tfrac{1}{2}\Delta z_\nu + \alpha z_\nu &= H_\nu^\varepsilon(x \cdot .D\zeta), \\ z_\nu|_{\partial\mathcal{O}} &= 0. \end{aligned}$$

By choosing conveniently the constant C in the definition of K_ε, one can check that T^ε maps K_ε into itself. Moreover, from (4.2), T^ε is continuous. Hence the fixed point property applies.

Now it is easy to check that H^ε satisfies all the assumptions of H, with the same constants, hence independent of ε. Therefore the solution u^ε of (4.74) remains bounded in $(W^{2,s}(\mathcal{O}))^N$ norm, independently of ε. Hence letting ε tend to 0, one obtains a solution of (4.65).

Proof of Theorem 4.1:
We can proceed on the solution u^α of (4.65) in $W^{2,s}$ which has been obtained, with the developments of sections 4.2, 4.3, 4.4, 4.5. Thanks to the assumptions of Theorem 4.1, we can obtain estimates in $W^{2,s}$ which are independent of α. We can then let α tend to 0, to complete the proof.

5. Hamiltonians Arising from Games

5.1. Setup: case (4.7), (4.8)

We define here the Hamiltonians $H_\nu(x,p)$ as follows:

$$H_\nu(x,p) = f_\nu(x) + g \cdot p_\nu + L_\nu(p) + \frac{\delta}{2}|p_\nu|^2, \tag{5.1}$$

where $L_\nu(p)$ has been defined by (3.8) and f_ν and g are the functions which arise in (2.10), (2.16). We first check (4.7), (4.8) with $\Gamma = I$. Using (3.53), (3.70), and formula (5.1), we obtain (4.8) with

$$\begin{aligned} \lambda_\nu^0 &= \frac{1-2\theta}{2\theta^2} + \frac{\delta}{2} + \frac{\varepsilon}{2}, \\ \lambda_\nu &= \|f_\nu\| + \frac{1}{2\varepsilon}\|g\|^2. \end{aligned} \tag{5.2}$$

Using (3.55) we have

$$\sum_\nu H_\nu(x,p) \geq -\|\sum_\nu f_\nu\| - \frac{N}{2\varepsilon}\|g\|^2 - \frac{\varepsilon}{2}|p|^2 + \frac{F_\delta(\theta)|p|^2}{(1-\theta)^2(1+(N-1)\theta)^2}.$$

We obtain (4.7), with

$$\lambda = \|\sum_\nu f_\nu\| + \frac{N}{2\varepsilon}\|g\|^2, \tag{5.3}$$

provided θ belongs to the validity intervals defined in (3.70). If the product $\lambda_\nu\lambda_\nu^0$ satisfies the condition (4.22), then the properties (4.7), (4.8) are satisfied.

Recalling that $F_\delta(\theta) \geq F(\theta)$, we may pick $\theta < \theta'$ or $\theta'' < \theta < \bar{\theta}$, and we get a condition of smallness on δ (note that $\theta < 0$), from (4.22).

Let us verify (4.11), (4.12), (4.13). We define the matrix Γ as

$$\begin{aligned} \Gamma_{\nu\mu} &= \delta_{\nu\mu} \quad \text{if} \quad \nu = 1,\dots N, \mu = 1,\dots,N-1, \\ \Gamma_{\nu N} &= -1 \quad \text{if} \quad \nu = 1,\dots N-1, \Gamma_{NN} = 1, \end{aligned}$$

hence setting $\tilde{p} = \Gamma p$, we have

$$\begin{aligned} H_\nu^\Gamma(x,\tilde{p}) &= H_\nu(x,p) - H_N(x,p), \quad \forall \nu < N, \\ H_N^\Gamma(x,\tilde{p}) &= H_N(x,p). \end{aligned}$$

Using (3.12) it follows, after easy computations, that

$$H_\nu^\Gamma(x,\tilde{p}) = Q(\tilde{p})\tilde{p}_\nu + H_\nu^0(x,\tilde{p}),$$

with

$$Q(\tilde{p}) = \frac{2\theta - 1}{(1-\theta^2(1+(N-1)\theta)} \sum_\mu \tilde{p}_\mu + \left(\frac{1-2\theta}{(1-\theta)^2} + \delta + N - 1\right) p_N,$$

$$H_\nu^0(x,\tilde{p}) = f_\nu - f_N + g \cdot \tilde{p}_\nu + \frac{1}{2}\left(\delta + \frac{1-2\theta}{(1-\theta)^2}\right) |\tilde{p}_\nu|^2, \ \nu < N,$$

$$H_N^0(x,\tilde{p}) = H_N(x,p) - Q(\tilde{p})\tilde{p}_N,$$

and the assumptions (4.11), (4.12), (4.13) are easily verified.

5.2. Setup: case (4.9), (4.10)

We are going to check (4.9), (4.10) with the following Γ:

$$\Gamma_{\nu\nu} = (N-1)\theta, \quad \Gamma_{\nu\mu} = -1, \quad \text{if} \quad \mu \neq \nu. \tag{5.4}$$

This matrix satisfies the maximum principle, provided $\theta > 1$. Indeed its inverse is

$$(\Gamma^{-1})_{\nu\nu} = \frac{(N-1)\theta - (N-2)}{(N-1)(\theta-1)((N-1)\theta+1)},$$

$$(\Gamma^{-1})_{\nu\mu} = \frac{1}{(N-1)(\theta-1)((N-1)\theta+1)}, \quad \mu \neq \nu,$$

which are positive, under the assumption on θ. This guarantees the maximum principle.

We set

$$\tilde{p} = \Gamma p,$$

and we have (see (3.45), (3.72))

$$M_\nu^\Gamma(\tilde{p}) = \tilde{M}_\nu(p).$$

According to formula (3.79) and recalling (3.44), we get

$$M_\nu^\Gamma(\tilde{p}) \geq -k_0|\tilde{p}_\nu|^2 \tag{5.5}$$

with

$$k_0 = \frac{1}{2(1-\theta)^2(1+(N-1)\theta)^2}\left\{\frac{(1-\theta)^2(1-\delta\theta(N-1))^2}{(N-1)\theta(1-\delta(\theta-1))} + (N-1)\theta(2\theta-1)\right.$$
$$\left. + \frac{\theta-1}{N-1} - \delta\left(\theta(\theta-1)(\theta(N-1)-N+3) + \frac{1}{N-1}\right)\right\}^+, \tag{5.6}$$

provided we assume

$$\theta - \frac{1}{2} - \frac{\delta}{2}(\theta-1)^2 > 0, \tag{5.7}$$

$$1 - \delta(\theta-1) > 0. \tag{5.8}$$

Next

$$H_\nu^\Gamma(x,\tilde{p}) = (N-1)\theta f_\nu(x) - \sum_{\mu\neq\nu} f_\mu(x) + g \cdot \tilde{p}_\nu + M_\nu^\Gamma(\tilde{p})$$
$$\geq -\lambda_\nu - \lambda_\nu^0|\tilde{p}_\nu|^2$$

with

$$\begin{aligned}\lambda_\nu &= \|(N-1)\theta f_\nu(x) - \sum_{\mu\neq\nu} f_\mu\| + \frac{1}{2\varepsilon}\|g\|^2,\\ \lambda_\nu^0 &= k_0 + \tfrac{\varepsilon}{2},\end{aligned} \tag{5.9}$$

and (4.10) is satisfied, provided $\lambda_\nu\lambda_\nu^0$ verifies (4.22). Next we have

$$\sum_\nu H_\nu(x,p) \le \|\sum_\nu f_\nu\| + \frac{N}{2\varepsilon}\|g\|^2 + \frac{\varepsilon}{2}|p|^2 + \frac{\hat{F}_\delta(\theta)|p|^2}{(1-\theta)^2(1+(N-1)\theta)^2},$$

and we obtain (4.9) provided the conditions of Lemma (3.5) are fulfilled, namely $\theta_0' < \bar{\theta}_\delta'$ and $\theta \in (\theta_{1\delta}''', \bar{\theta}_\delta')$. Since $\theta > 1$, and thanks to (3.66), this reduces to $\bar{\theta}_\delta' > 1$, and $\theta \in (1, \bar{\theta}_\delta')$. Considering (3.61), this means

$$\delta < \frac{1+2N(N-1)}{N^2}, \tag{5.10}$$

$$1 < \theta < \frac{1}{\delta}\left(1 - \frac{\delta}{N-1} + \sqrt{1 + \frac{\delta}{(N-1)^2}}\right). \tag{5.11}$$

Summing up we can assert the

Proposition 5.1. *For the Hamiltonians* (5.1), *the assumptions* (4.11), (4.12), (4.13), *are satisfied. The conditions* (4.7), (4.8) *are verified with* $\lambda_\nu^0, \lambda_\nu$ *given by* (5.2), λ *given by* (5.3), *with* $\Gamma = I$ *provided* θ *belongs to the validity intervals defined in* (3.70) *and* $\lambda_\nu\lambda_\nu^0$ *verifies the condition* (4.22). *On the other hand, the conditions* (4.9), (4.10) *are satisfied with* $\lambda_\nu^0, \lambda_\nu$ *given by* (5.9), λ *given by* (5.3), *with* Γ *defined by* (5.4), *provided* $\theta > 1$ *and* δ *verifies* (5.7), (5.8), (5.10), (5.11), $\lambda_\nu^0\lambda_\nu$ *verifies the condition* (4.22).

6. Solution of the Stochastic Game Problem

6.1. Bellman system

We relate to problem (2.17) the system of Bellman equations

$$\begin{aligned}-\tfrac{1}{2}\Delta u_\nu &= f_\nu + g\cdot Du_\nu + L_\nu(Du) + \tfrac{\delta}{2}|Du_\nu|^2,\\ u_\nu|_{\partial\mathcal{O}} &= 0,\end{aligned} \tag{6.1}$$

and we assume that all conditions of Proposition 5.1 are satisfied. Therefore, according to Theorem 4.2, there exists a solution in $(W^{2,s}(\mathcal{O}) \cap W_0^{1,s}(\mathcal{O}))^N, \forall 2 \le s < \infty$. We define the feedbacks

$$\hat{v}_\nu(x) = v_\nu(Du(x)), \tag{6.2}$$

where $v_\nu(p)$ are given by (3.6). These functions are bounded and continuous. To such a feedback we associate the stochastic processes

$$\hat{v}_\nu(t) = \hat{v}_\nu(x(t)), \tag{6.3}$$

where $x(t)$ is the process defined by (2.4). Our objective is to check that such a vector process $\hat{v}(t)$ is the Nash point of the functionals $J_\nu(x, v(\cdot))$ defined by (2.16), in the sense of (2.17).

We state the

Theorem 6.1. *We assume all the conditions of Proposition* 5.1, *so that there exists a solution of the system* (6.1), *in the space* $(W^{2,s}(\mathcal{O}) \cap W_0^{1,s}(\mathcal{O}))^N$. *Then the control defined by* (6.3), *corresponding to the feedback defined by* (6.2) *is a Nash point of the functionals* $J_\nu(x, v(\cdot))$ *defined by* (2.16), *in the sense of* (2.17). *Moreover one has*

$$u_\nu(x) = J_\nu(x, \hat{v}(\cdot)). \tag{6.4}$$

6.2. Proof of Theorem 6.1.

Note that from the definition (6.2), we have

$$\begin{aligned} L_\nu(Du) &= \tfrac{1}{2}|\hat{v}_\nu(Du)|^2 + \theta \hat{v}_\nu(Du) \cdot \bar{\hat{v}}_\nu(Du), \\ &\leq \tfrac{1}{2}|v_\nu|^2 + \theta v_\nu \cdot \bar{\hat{v}}_\nu(Du), \quad \forall x, \forall v_\nu. \end{aligned} \tag{6.5}$$

Applying this inequality to $x = x(t)$, yields, from (6.3)

$$\begin{aligned} L_\nu(Du(x(t))) &= \frac{1}{2}|\hat{v}_\nu(t)|^2 + \theta \hat{v}_\nu(t) \cdot \bar{\hat{v}}_\nu(t) \\ &\leq \frac{1}{2}|v_\nu(t)|^2 + \theta v_\nu(t) \cdot \bar{\hat{v}}_\nu(t), \end{aligned} \tag{6.6}$$

for any control v_ν.

Consider the probability $P_{x,\hat{v}}$ defined by (2.8) where $\hat{v}(\cdot)$ is the process defined by (6.3). Note that for the system $(\Omega, \mathcal{A}, \mathcal{F}^t, P_{x,\hat{v}}, w_{x,\hat{v}})$ we have (see (2.10))

$$dx = (g(x(t)) + \sum_\mu \hat{v}_\mu(t))dt + dw_{x,\hat{v}}(t), \quad x(0) = x. \tag{6.7}$$

From Ito's formula we have

$$\begin{aligned} du_\nu(x(t)) = \Bigg(& Du_\nu(x(t)) \cdot (g(x(t)) + \sum_\mu \hat{v}_\mu(t)) \\ & + \frac{1}{2}\Delta u_\nu(x(t)) \Bigg) dt + Du_\nu(x(t)) dw_{x,\hat{v}}(t), \end{aligned}$$

and from Bellman equation (6.1), we deduce

$$\begin{aligned} du_\nu(x(t)) = \Bigg(& -f_\nu(x(t)) - \frac{1}{2}|\hat{v}_\nu(t)|^2 - \theta \hat{v}_\nu(t) \cdot \bar{\hat{v}}_\nu(t) \\ & - \frac{\delta}{2}|Du_\nu(x(t))|^2 \Bigg) dt + Du_\nu(x(t)) \cdot dw_{x,\hat{v}}(t). \end{aligned} \tag{6.8}$$

Integrating between 0 and τ_x yields

$$\begin{aligned} u_\nu(x) = & \int_0^{\tau_x} [f_\nu(x(t)) + \frac{1}{2}|\hat{v}_\nu(t)|^2 + \theta\hat{v}_\nu(t)\cdot\bar{\hat{v}}_\nu(t)]dt \\ & - \int_0^{\tau_x} Du_\nu(x(t))\cdot dw_{x,\hat{v}}(t) + \frac{\delta}{2}\int_0^{\tau_x} |Du_\nu(x(t)|^2 dt. \end{aligned} \tag{6.9}$$

Hence

$$\begin{aligned} & \exp\delta\int_0^{\tau_x} [f_\nu(x(t)) + \frac{1}{2}|\hat{v}_\nu(t)|^2 + \theta\hat{v}_\nu(t)\cdot\bar{\hat{v}}_\nu(t)]dt \\ & = \exp\delta u_\nu(x)\exp\left(\int_0^{\tau_x} \delta Du_\nu(x(t))dw_{x,\hat{v}}(t) - \frac{1}{2}\int_0^{\tau_x} |\delta Du_\nu(x(t))|^2 dt\right). \end{aligned} \tag{6.10}$$

Since

$$\exp\left(\int_0^t \delta Du_\nu(x(s))dw_{x,\hat{v}}(s) - \frac{1}{2}\int_0^t |\delta Du_\nu(x(s))|^2 ds\right)$$

is an $\mathcal{F}^t, P_{x,\hat{v}}$ martingale, and τ_x is a stopping time, we deduce the formula

$$\exp\delta u_\nu(x) = E_{x,\hat{v}}\exp\delta\int_0^{\tau_x} [f_\nu(x(t)) + \frac{1}{2}|\hat{v}_\nu(t)|^2 + \theta\hat{v}_\nu(t)\cdot\bar{\hat{v}}_\nu(t)]dt, \tag{6.11}$$

hence (6.4) is demonstrated.

Next considering an arbitrary control $v(\cdot)$, we manufacture the control $(v_\nu(\cdot), \hat{v}^\nu(\cdot))$, in which we take for all components $\mu \neq \nu$ the control $\hat{v}_\mu(\cdot)$ defined by (6.3), and for the component ν the control $v_\nu(\cdot)$. Performing similar calculations as above, and taking account this time of the inequality (6.6), we can check

$$u_\nu \leq J_\nu(x, v_\nu, \hat{v}^\nu)$$

which establishes the inequality (2.17).

The proof has been completed. □

7. Appendix 1: Discussion on the Smallness of $\lambda_\nu\lambda_\nu^0$

First we argue that a smallness condition on $\lambda_\nu\lambda_\nu^0$ (see (4.8) and (4.10)) cannot be avoided. Consider indeed the case of $N = 1$, and in dimension 1, the equation

$$-\frac{1}{2}u'' = \lambda + \lambda^0 u'^2,$$

$$u(0) = u(1) = 1.$$

For $\lambda^0 \leq 0$ it is easy to check that there exists a unique bounded solution. For $\lambda^0 > 0$ the unique possible solution is

$$u(x) = \frac{1}{2\lambda^0}\log|\frac{\cos\sqrt{\lambda^0\lambda}(2x-1)}{\cos\sqrt{\lambda^0\lambda}}|,$$

which is bounded if and only if $\lambda^0\lambda < \dfrac{\pi^2}{4}$.
Let us compare with the limitation obtained in (4.22). In the present example, the Green function $G^\xi(x)$ is given by:

$$G^\xi(x) = 2(1-\xi)x \quad 0 \leq x \leq \xi,$$
$$2\xi(1-x) \quad \xi \leq x \leq 1$$

and

$$\int_0^1 G^\xi(x)dx = \xi(1-\xi) \leq \frac{1}{4}.$$

Therefore, condition (4.22) means here $\lambda^0\lambda < 1$. So we do not get the optimal limitation by this method, as can be expected.

So the issue of the improvement of the limitation on $\lambda^0\lambda$ is an interesting question. We shall check in the following result that an improvement can arise from the presence of an adequate drift g. This is reminiscent of what occurs in the finite horizon problem (see [6]).

A variant of Theorem 4.1

We consider here the problem

$$\begin{aligned} -\tfrac{1}{2}\Delta u_\nu - g \cdot Du_\nu &= H_\nu(x \cdot Du), \\ u_\nu|_{\partial\mathcal{O}} &= 0, \end{aligned} \tag{7.1}$$

which, compared to (4.1) has a linear first order term $-g \cdot Du_\nu$. Of course, the usual treatment would be to incorporate it in the Hamiltonian H_ν, but as far as condition (4.8) is concerned, this implies a deterioration of the smallness condition on $\lambda_\nu\lambda_\nu^0$. In fact, we shall see here that this term can relax to a large extent the limitation on $\lambda_\nu\lambda_\nu^0$. provided div $g > 0$.

We proceed as follows. Considering the function (4.20), E_ν, we can check that

$$-\frac{1}{2}\Delta E_\nu - g \cdot DE_\nu \leq 2\lambda_\nu\lambda_\nu^0 E_\nu. \tag{7.2}$$

Since everything relates to the ν equation, we shall drop the symbol ν in the sequel. So, also

$$\begin{aligned} -\tfrac{1}{2}\Delta(E-1) - g \cdot D(E-1) &\leq 2\lambda\lambda^0(E-1) + 2\lambda\lambda^0, \\ E - 1|_{\partial\mathcal{O}} &= 0. \end{aligned} \tag{7.3}$$

We first check that under some smallness conditions, we can obtain an estimate on $E-1$ in $L^2(\mathcal{O})$. Indeed, testing (7.3) with $(E-1)^+$ yields

$$\frac{1}{2}\int_{\mathcal{O}} |D(E-1)^+|^2 dx + \int_{\mathcal{O}} \frac{div\ g}{2}(E-1)^{+2}$$
$$\leq \lambda\lambda^0 \int_{\mathcal{O}} (E-1)^{+2} dx + 2\lambda\lambda^0 \int_{\mathcal{O}} (E-1)^+ dx.$$

From Poincaré's inequality, we get

$$\int_{\mathcal{O}}(k_0+div\ g)(E-1)^{+2}\leq 4\lambda\lambda^0\int_{\mathcal{O}}(E-1)^{+2}dx+4\lambda\lambda^0\int_{\mathcal{O}}(E-1)^{+}dx,$$

and if the smallness condition

$$4\lambda\lambda^0<k_0+inf\ div\ g \tag{7.4}$$

is satisfied, then clearly

$$\int_{\mathcal{O}}(E-1)^2dx\leq|\mathcal{O}|+\int_{\mathcal{O}}(E-1)^{+2}dx\leq C, \tag{7.5}$$

hence also

$$\int_{\mathcal{O}}E^2dx\leq C. \tag{7.6}$$

From that knowledge, we are going to check that E is bounded in L^∞, without using anymore any smallness condition. We test (7.3) with EG^ξ, using again the Green function (4.15) (although the Green function related to the operator $-\frac{1}{2}\Delta-g\cdot D$ is also possible). We obtain

$$\begin{aligned}&\frac{1}{2}\int_{\mathcal{O}}|DE|^2G^\xi dx+\frac{1}{4}\int_{\mathcal{O}}D(E^2-1)DG^\xi dx\\&\leq\int_{\mathcal{O}}E^2G^\xi(2\lambda\lambda^0-\frac{1}{2}div\ g)dx\\&-\int_{\mathcal{O}}\frac{1}{2}E^2g\cdot\ \text{grad}\ G^\xi dx+\frac{1}{2}\int_{\mathcal{O}}(G^\xi div\ g+g\cdot\ \text{grad}\ G^\xi)dx,\end{aligned} \tag{7.7}$$

and from the definition of the Green function we deduce

$$\begin{aligned}E^2(\xi)\leq 1+\int_{\mathcal{O}}E^2G^\xi(4\lambda\lambda^0-div\ g)dx-\int_{\mathcal{O}}E^2g\cdot\ \text{grad}\ G^\xi dx\\+\int_{\mathcal{O}}(G^\xi div\ g+g\cdot\ \text{grad}\ G^\xi)dx.\end{aligned} \tag{7.8}$$

We now make use of (7.6). Consider the set $\{x:E(x)>L\}$, where L is large >1. We split the integrals on the right-hand side of (7.8) into the integral on this set and on its complement. Assuming ξ is chosen to be the maximum of E, supposed to be larger than 1 (otherwise the L^∞ bound is obvious), we can then deduce from (7.8) the inequality

$$\begin{aligned}&\|E^2\|_{L^\infty}\leq 1+L^2(4\lambda\lambda^0+\|div\ g\|)\int_{\mathcal{O}}G^\xi dx\\&+L^2\|g\|\int_{\mathcal{O}}|DG^\xi|dx+\|E^2\|_\infty(4\lambda\lambda^0+\|div\ g\|)\int_{\{E>L\}}G^\xi dx\\&+\|E\|_\infty\|g\|\int_{\{E>L\}}|DG^\xi|dx+C.\end{aligned} \tag{7.9}$$

Now from (7.6) we have

$$\text{Meas}\{E>L\}\leq\frac{C}{L^2}, \tag{7.10}$$

and from the estimates (4.16), (4.17), we can assert, thanks to Hölder's inequality,

$$\int_{\{E>L\}} G^\xi dx \leq C\|G^\xi\|_{L^q}\frac{1}{L^{2q'}}, \tag{7.11}$$

$$\int_{\{E>L\}} |DG^\xi| dx \leq C\|G^\xi\|_{L^r}\frac{1}{L^{2r'}}. \tag{7.12}$$

By picking L sufficiently large, the coefficient in front of $\|E^2\|_\infty$ in the right-hand side of (7.9) can be made as small as we wish, hence strictly smaller than 1. Then (7.9) yields an estimate on the L^∞ norm of E.

The rest of the proof of Theorem 4.2 is unchanged. □

Considering the limitation (7.4), it can be very good if $\inf div\ g$ is very large. However this estimate, which can be also used when $g = 0$, is not very good in that case. For example in our one dimensional-example, the Poincaré constant $k_0 = 1$, hence $\lambda\lambda^0 < \frac{1}{4}$, which is much worse than 1 obtained from (4.22).

8. Appendix 2: Global smallness condition

If we consider (4.7), (4.8) with $\Gamma = I$, we get the property

$$-\lambda - \sum_{\mu\neq\nu}\lambda_\mu - \sum_{\mu\neq\nu}\lambda^0_\mu|p_\mu|^2 \leq H_\nu(x,p) \leq \lambda_\nu + \lambda^0_\nu|p_\nu|^2 \tag{8.1}$$

and $\lambda_\nu\lambda^0_\nu$ not too large. This is reminiscent of, although *not equivalent* to, a global smallness assumption on k, K where

$$|H(x,p)| \leq k|p|^2 + K. \tag{8.2}$$

In that case things simplify greatly (see [8]).

For the sake of completeness we sketch the main arguments in the case of a global small assumption. First to obtain L^∞ estimates, we test the equation

$$-\frac{1}{2}\Delta u_\nu = H_\nu(x, Du) \tag{8.3}$$

with $G^\xi u_\nu$, which yields

$$\int_{\mathcal{O}} D|u|^2 DG^\xi dx + \frac{1}{2}\int_{\mathcal{O}} G^\xi|Du|^2 dx = \int_{\mathcal{O}} G^\xi|u||H(x,Du)|dx.$$

Hence if ξ is a point where $|u|$ reaches its maximum, we have

$$\|u\|^2_\infty + \frac{1}{2}\int_{\mathcal{O}} G^\xi|Du|^2 dx \leq \|u\|_\infty\int_{\mathcal{O}} G^\xi(K + k|Du|^2)dx,$$

thus if we have

$$2k\|u\|_\infty < 1, \tag{8.4}$$

then it follows that

$$\|u\|_\infty < K \int_{\mathcal{O}} G^\xi dx.$$

So it is sufficient to assume

$$2kK\|G\|_{L^1} < 1. \tag{8.5}$$

The H_0^1 estimate follows, since testing (8.3) with u_ν yields

$$\frac{1}{2}\int_{\mathcal{O}} |Du|^2 dx \leq \|u\|_\infty \int_{\mathcal{O}} |H| dx$$
$$\leq \|u\|_\infty \Big(\int_{\mathcal{O}} k|Du|^2 dx + K \text{ Meas } \mathcal{O}\Big),$$

and we make use of (8.4) to derive the H_0^1 estimate. To obtain the C^δ and $W^{1,p}$ estimate, $2 \leq p < 2+\varepsilon$, we test with $(u_\nu - C_\nu^R)\tau_R^2$, where τ_R, C_ν^R have been defined in (4.44), (4.45). We obtain

$$\frac{1}{2}\int_{\mathcal{O}} |Du|^2\tau_R^2 + \int_{\mathcal{O}} (u_\nu - c_\nu^R) Du_\nu D\tau_R \tau_R dx \leq \|u - c^R\|_\infty \int_{\mathcal{O}} (k|Du|^2 + K)\tau_R^2 dx$$
$$\leq 2\|u\|_\infty \int_{\mathcal{O}} (k|Du|^2 + K)\tau_R^2 dx,$$

so if we assume a more stringent assumption than (8.5), namely

$$4kK\|G\|_{L^1} < 1, \tag{8.6}$$

which implies

$$4k\|u\|_\infty < 1,$$

then we get

$$c_0 \int_{B_R} |Du|^2 dx \leq C \int_{B_{2R}} |Du| \frac{|u - c^R|}{R} dx + CR^n,$$

which is the condition (4.46).

Similarly we can obtain (4.51), by testing with $(u_\nu - c_\nu^R)G^{x_0}\tau_R^2$, where G^{x_0} has been defined in (4.53). We get this time

$$\frac{1}{2}\int_{\mathcal{O}} |Du|^2 G^{x_0}\tau_R^2 dx + \int_{\mathcal{O}} (u_\nu - C_\nu^R) Du_\nu D\tau_R \tau_R G^{x_0} dx$$
$$\leq 2\|u\|_\infty \int_{\mathcal{O}} (k|Du|^2 + K) G^{x_0}\tau_R^2 dx,$$

and by virtue of (8.6),

$$c_0 \int_{\mathcal{O}} |Du|^2 G^{x_0}\tau_R^2 dx \leq \int_{\mathcal{O}} |u - C^R||Du||D\tau_R|\tau_R G^{x_0} dx$$
$$+ C\int_{\mathcal{O}} G^{x_0}\tau_R^2 dx.$$

Then making use of the properties of the Green function (see (4.54)), using Hölder's inequality then Poincaré's inequality, we derive (4.51) easily. □

It is also possible to get rid of the more stringent assumption (8.6) and to assume only (8.5), see [8].

References

[1] J.P. AUBIN, *Mathematical Methods of Game and Economic Theory*, Studies in Mathematics and its Applications, North Holland, Amsterdam, Holland (1976).

[2] A. BENSOUSSAN, J. FREHSE, *Topics on Nonlinear Partial Differential Equations and Applications*, to be published, Springer Verlag.

[3] A. BENSOUSSAN, J. FREHSE *Nonlinear Elliptic Systems in Stochastic Game Theory*, Journal fur Die Reine und angewandte Mathematik, Vol 305, pp. 23–67 (1984).

[4] A. BENSOUSSAN, J. FREHSE *Stochastic Games for N-Players*, to be published in the book in honor of Professor D.G. Luenberger.

[5] A. BENSOUSSAN, J. FREHSE, H. NAGAI *Some results on risk sensitive control with full observation*, Applied Mathematics and Optimization 37 (1998), (pp. 1–41).

[6] A. BENSOUSSAN, H. NAGAI *Conditions for no Breakdown and Bellman equation Equations of Risk Sensitive control*, Applied Mathematics and Optimization, to appear.

[7] F.W. GEHRING *The L^p integrability of the partial derivatives of a quasi conformal mapping*, Acta Mathematica, Vol.130, (1973), pp. 266–277.

[8] S. HILDEBRANDT, K.O. WIDMAN *On the Hölder continuity of weak solutions of quasi-linear elliptic systems of second order*, Ann. Sc. Norm. Sup Pisa, 4 (1977), pp145–178.

[9] L. NIRENBERG *An extended interpolation inequality*, Ann. Sc. Norm. Sup Pisa, 20 (1966), pp 733–737.

[10] H. NAGAI *Bellman equations of risk-sensitive control*, SIAM J. Control Optim. 34, 1 (1966), pp 74–101.

[11] J. NASH *Equilibrium points in n-persons Games*, Proc. of the National Academy of Sciences of the USA, vol. 36, pp. 48–49 (1950).

[12] K.O. WIDMAN *Hölder continuity of solutions of elliptic equations*, Manuscripta Math. 5 (1971), pp. 299–308.

University of Paris-Dauphine and CNES
2, place Maurice Quentin,
75775 Paris Cedex 16
France

Universität Bonn
Institut für Angewandte Mathematik
Wegelerstr. 6
53115 Bonn
Germany

Progress in Probability, Vol. 53, 67–90

An Overall View of Stochastics in Colombeau Related Algebras

Uluğ Çapar and Hüseyin Aktuğlu

Abstract. Stochastic analysis in Colombeau algebras was initiated by the works of Çapar, Oberguggenberger and Russo in the early 1990s. The present communiqué is a comparative and critical account of the major research papers that have appeared so far in this particular field of study. It turns out that application of the nonlinear distribution theories of Colombeau and Rosinger to stochastic analysis provides new horizons in dealing with singular objects like nonlinear functions of white noise, Wiener functionals and nonlinear stochastic PDE, especially when the initial data is irregular or distributional.

1. Introduction and Preliminaries

In the non-linear theories of generalized functions (g.f.'s) introduced by Colombeau and Rosinger ([3], [4], [21], [22]) in the 1980s, the associative and commutative differential algebra into which $\mathcal{D}'$ (or S') is linearly imbedded, is usually constructed as a factor space of the form

$$\mathcal{G}(\Omega) = \mathcal{E}_M(\Omega)/\mathcal{N}(\Omega) \tag{1.1}$$

where Ω is an open subset of $\mathcal{R}^n$ and $\mathcal{E}_M$ is a subalgebra of the sequence space $\mathcal{E}(\Omega) = (\mathcal{C}^\infty(\Omega))^{\mathcal{A}}$ consisting of moderate elements, $\mathcal{N}$ is the ideal of null elements and $\mathcal{A}$ is an index set.

In works surveyed by this article, different index set systems have been used:

A) $\mathcal{A} = N$ (cf. [17]).

B) $\mathcal{A} = \{\phi \in S(\mathcal{R}^n) : \int_{\mathcal{R}^n} \phi(x)dx = 1, \int_{\mathcal{R}^n} x^j \phi(x)dx = 0,\ j \in N_o^n,\ |j| \geq 1\}\}$, i.e., rapidly decreasing smooth functions on $\mathcal{R}^n$ with a compactly supported Fourier transform which coincides with constant function 1 around 0, (c.f. [13]).

C) $\mathcal{A}_q = \{\phi \in \mathcal{D}(\mathcal{R}^n) : \int_{\mathcal{R}^n} \phi(x)dx = 1, j \in N_0^n,\ 1 \leq |j| \leq q\}$,
$\mathcal{A} = \mathcal{A}_1 \supset \mathcal{A}_2 \supset \cdots,\ \ \cap_{q=1}^{\infty} \mathcal{A}_q = \emptyset$ (cf. [20], [21]).

D) $\bar{\mathcal{A}}_0(\mathcal{R}) = \{\phi_1 \in \mathcal{D}(\mathcal{R}) : \phi_1$ is even, constant in a 0-nbd and $\int_0^\infty \phi_1(x)dx = \frac{1}{2}\}$,
$\bar{\mathcal{A}}_q(\mathcal{R}) = \{\phi_1 \in \bar{\mathcal{A}}_0 : \int_0^\infty x^{\frac{j}{m}} \phi_1(x)dx = 0,\ \ 1 \leq j \leq q,\ 1 \leq m \leq q\}$,
$q = 1, 2, \ldots$, and
$\bar{\mathcal{A}}_q(\mathcal{R}^n) = \{\phi \in \mathcal{D}(\mathcal{R}^n) : \exists \phi_1 \in \bar{\mathcal{A}}_q(\mathcal{R}) \ni\ \phi(x) = \frac{2n}{\Omega_n} \phi_1(\|x\|^n)\}$,

$q = 0, 1, 2, \ldots$, where Ω_n is the surface area of an $n-1$ dimensional sphere of radius 1, (c.f. [2]). We have $\mathcal{A} = \bar{A}_0 \supset \bar{A}_0 \supset \bar{A}_1 \supset \cdots \supset \bar{A}_q \cdots$; $\cap_{q=0}^{\infty} \bar{A}_q = \emptyset$.

Each index set has its own merits and disadvantages. With $\mathcal{A}$ in A) there is no privileged (canonical) imbedding of $\mathcal{E}'(\Omega)$ (distributions of compact support) into $\mathcal{G}(\Omega)$. But in return, in this cardinality the measurability problems confronted in the stochastic case can be more favorably tackled.

Index sets in B), C) and D) are closed under the formation of mollifying sequences, e.g., if $\phi \in \mathcal{A}$, then $\phi_\epsilon(x) := \frac{1}{\epsilon^n}\phi(\frac{x}{\epsilon})$, $x \in \mathcal{R}^n$, is also in $\mathcal{A}$ for all $\epsilon > 0$.

In B), the use of $S(\mathcal{R}^n)$ allows us to construct $\mathcal{A}$ consisting of functions with all moments vanishing, thus avoiding the dependence of analyses on index q as in C) and D), but in return even in the case $\Omega = \mathcal{R}^n$, one has to consider the imbedding of compactly supported distributions first and extend it to $\mathcal{D}'$ by the localization (sheaf property) through an ad hoc cut-off process.

The index system given in C) is the most natural one given in the standard texts by Colombeau and Rosinger (c.f. [3], [4], [21], [22]).

A little more complicated contruction proposed by Biagioni ([2]) has the advantage of being independent of the choice of orthonormal bases and in many instances of the dimension (e.g., a unique $\bar{\mathcal{C}}$ of generalized constants for all n).

$\mathcal{E}_M(\Omega)$ consists of sequences $\{R(\phi)\}_{\phi\in\mathcal{A}}$ of smooth functions which grow at most polynomially in $\frac{1}{\epsilon}$ as $\epsilon \to 0$ when evaluated at ϕ_ϵ , uniformly in compact sets along with derivatives of all orders. In the case of an index set in A), one has to replace $\frac{1}{\epsilon}$ with ν and $\phi_\nu(x) = \nu^n\phi(\nu x)$). Thus for instance with the index set of C) (similarly for D)):

$$\begin{aligned} \mathcal{E}_M(\Omega) = \quad & \{R \in \mathcal{E}(\Omega) : \forall K \subset\subset \Omega,\ \forall\alpha \in N_0^n, \exists N \geq 0 \ni \forall\phi \in \mathcal{A}_N, \\ & \|D^\alpha R(\phi_\epsilon, x)\|_{L^\infty(K)} = O(\epsilon^{-N}) \text{ as } \epsilon \to \infty\}. \end{aligned} \tag{1.2}$$

With index sets in A) or B) it takes a simpler form, e.g., for $\mathcal{A}$ in A):

$$\begin{aligned} \mathcal{E}_M(\Omega) = \quad & \{R \in \mathcal{E}(\Omega) : \forall K \subset\subset \Omega,\ \forall\alpha \in N_0^n, \exists N > 0 \ni \\ & \|D^\alpha R_\nu(x)\|_{L^\infty(K)} = O(\nu^{-N}) \text{ as } \nu \to \infty\}. \end{aligned} \tag{1.3}$$

The ideal of null elements is given as (for C)) :

$$\begin{aligned} \mathcal{N}(\Omega) = \{R \in \mathcal{E}(\Omega) : \forall K \subset\subset \Omega, \forall\alpha \in N_0^n, \exists N \geq 0 \ni \forall q \geq N, \forall\phi \in \mathcal{A}_q; \\ \|D^\alpha R(\phi_\epsilon)\|_{L^\infty(K)} = O(\epsilon^{q-N}) \text{ as } \epsilon \to 0\}, \end{aligned} \tag{1.4}$$

and for A):

$$\begin{aligned} \mathcal{N}(\Omega) = \quad & \{R \in \mathcal{E}(\Omega) : \forall K \subset\subset \Omega, \forall\alpha \in N_0^n, \forall N > 0, \\ & \|D^\alpha R_\nu(x)\|_{L^\infty(K)} = O(\nu^{-N}) \text{ as } \nu \to \infty\}. \end{aligned} \tag{1.5}$$

Generalized functions, i.e., members of the factor space

$$\mathcal{G}(\Omega) = \mathcal{E}_M(\Omega)/\mathcal{N}(\Omega),$$

are of the form $T = R_T + \mathcal{N}$, $R_T \in \mathcal{E}_M$, e.g.,

- $\delta = R_\delta + \mathcal{N}$ (Dirac delta function), $R_\delta(\phi, x) = \phi(-x)$, $\forall \phi \in \mathcal{A}(\mathcal{R})$.
- $H = R_H + \mathcal{N}$ (Heaviside function), $R_H(\phi, x) = \int_{-x}^{\infty} \phi(t)dt$, $\forall \phi \in \mathcal{A}(\mathcal{R})$.
- $x^m.\delta = g + \mathcal{N} \in \mathcal{G}(\mathcal{R})$, $g(\phi, x) = x^m \phi(-x) \notin \mathcal{N}(\mathcal{R})$. etc.

Differentiation and algebraic operations are performed on the representatives. The (Schwartz) distributions are imbedded linearly into $\mathcal{G}(\Omega)$ via convolution: $T \to f_T + \mathcal{N}$, where

$$f_T(\phi) = T * \phi|_\Omega, \text{ i.e., } f_T(\phi, x) = T(\check{\tau}_x \phi) = T_y(\phi(x - y)); \tag{1.6}$$

in fact,

$$f_T(\phi, x) = T(\tau_x \phi) = T_y(\phi(y - x))$$

would also be possible, (c.f. [21], pp. 62), $f_T \in \mathcal{E}_M(\Omega)$.

This imbedding respects distributional derivatives and renders $\mathcal{C}^\infty(\Omega)$ as a faithful subalgebra. If we use index functions in B), the imbedding has to start with a localization to compact supports, then using (1.6), it can be extended to $\mathcal{D}'(\Omega)$. This is achieved by fixing an expanding sequence $\{K_m\}(m = 1, 2, \cdots)$ of compact subsets covering Ω, one being included in the interior of the next one and using cut-off functions $\chi_m, (m = 1, 2, \cdots)$. For any m, the smooth function χ_m is supported in K°_{m+1} and takes value 1 on K_m. Then for $T \in \mathcal{D}'(\Omega)$, the equivalence classes of $f_m(\phi, x) = [(\chi_m T) * \phi](x)$, $(\phi \in S(\mathcal{R}^n), x \in \Omega)$ define $T_m \in \mathcal{G}(K^\circ_m)$ satisfying the coherence property. With this sequence we can define a generalized function $i(T) \in \mathcal{G}(\Omega)$ such that $i(T)|_{K^\circ_m} = T_m, \forall m$. This construction is independent of the choices of K_m and χ_m.

In case the index sets consist of functions with compact support as in C) and D) and if $\Omega = \mathcal{R}^n$, then (1.6) yields a canonical imbedding of $\mathcal{D}'(\mathcal{R}^n)$ into $\mathcal{G}(\mathcal{R}^n)$.

If Ω is a proper open subset of $\mathcal{R}^n$ and the index sets are as in C) and D), a cut-off function technique similar to the one given above can be used. As indicated by Biagioni ([2], pp.18) the imbedding which results will not be essentially different from that of the canonical imbedding when $\Omega = \mathcal{R}^n$.

In the case of $\mathcal{A} = N$, any $\phi \in S(\mathcal{R}^n)$ that has vanishing moments generates a mollifying sequence $\phi_\nu(x) = \nu^n \phi(\nu x)$ and an imbedding of $\mathcal{E}'(\Omega)$ through the equivalence class of $\{T * \phi_\nu|_\Omega\}_{\nu \in N}$, $T \in \mathcal{E}'(\Omega)$. Then the extension to an imbedding of all $\mathcal{D}'(\Omega)$ follows from the localization property.

For the index set in D) there is a bijection $I^m_n : \bar{\mathcal{A}}_0(\mathcal{R}^n) \to \mathcal{A}_0(\mathcal{R}^n)$ $(m, n = 1, 2, \ldots)$ with properties:

$$\begin{aligned} &(1) \quad I^m_n(\bar{\mathcal{A}}_q(\mathcal{R}^m)) = \bar{\mathcal{A}}_q(\mathcal{R}^n), \ q = 1, 2, 3, \ldots, \\ &(2) \quad I^m_m = i_d, \\ &(3) \quad I^m_n(\phi_\epsilon) = (I^m_n \phi)_{\epsilon^{\frac{m}{n}}} = I^m_n, \ p = 1, 2, 3, \ldots, \\ &(4) \quad I^p_n \circ I^m_p = I^m_n, \quad p = 1, 2, 3, \ldots. \end{aligned} \tag{1.7}$$

In the sequel we are going to use the index sets in D) unless otherwise stated.

The algebra of generalized constants (g.c.) $\bar{\mathcal{C}} = \mathcal{E}^\circ_M / \mathcal{J}$ is the algebra of g.f.'s on $\mathcal{R}^\circ = \{0\}$. For $T \in \mathcal{G}(\Omega), T = f_T + \mathcal{N}$, the point value of $T(x)$, $x \in \mathcal{R}^n$ is given by $T(x) = f_{T,x} + \mathcal{J}$, $f_{T,x}(\phi) = f_T(\phi, x), \phi \in \mathcal{A}(\mathcal{R})$. If $\bar{z} \in \bar{\mathcal{C}}$, it is said that $\bar{z}$ is associated to complex number z, denoted $\bar{z} \vdash z$, if there is a representation $\bar{z} = h + \mathcal{J} \in \bar{\mathcal{C}}$, such that $\exists q \in N^+, \forall \phi \in \mathcal{A}_q \ \lim_{\epsilon\downarrow 0} h(\phi_\epsilon) = z$. It turns out that $\delta(0)$ has no associated complex number, on the other hand g.c. $\bar{w} = g + \mathcal{J}$ where $g(x) = \int_{\mathcal{R}} |x|\phi(x)dx, \phi \in \mathcal{A}(\mathcal{R})$ satisfies $\bar{w} \vdash 0$, nevertheless $\bar{w} \neq \mathcal{J}$. For a linear distribution $T \in \mathcal{D}'$, the integrals $\int(\psi.T)(x)dx, (\psi \in \mathcal{D})$ give a local (therefore global) characterization of T. The counterpart of this integral for $T = f_T + \mathcal{N} \in \mathcal{G}(\Omega)$ will be the following g.c.:

$$\int (\psi \odot T)(x)dx = h_\psi + \mathcal{J},\ h_\psi(\phi) = \int_\Omega f_T(\phi, x)\psi(x)dx,\ \psi \in \mathcal{D}(\Omega),$$

($\odot$ denotes multiplication in $\mathcal{G}$). However these g.c.'s do not characterize T. If it happens that for a representative f_T we have

$$\lim_{\epsilon \to 0} \int f_T(\phi_\epsilon, x)\psi(x)dx = V(\psi)$$

for some $V \in \mathcal{D}'(\Omega)$ (i.e., $h_\psi \vdash V(\psi)$), we say that T is associated with Schwartz distribution V. This association is denoted by $T \approx V$.

If $\int(\psi \odot T)(x)dx = \mathcal{J}, \forall\psi \in \mathcal{D}$, then T is said to be test null and this is indicated by $T \sim 0$. Then the following equivalence relations are defined in $\mathcal{G}(\Omega)$:

$$\begin{aligned} T_1 \approx T_2 &\iff T_1 - T_2 \approx 0, \\ T_1 \sim T_2 &\iff T_1 - T_2 \sim 0. \end{aligned} \tag{1.8}$$

For instance $H^q \approx H,\ H \odot \delta \approx \frac{1}{2}\delta$, etc. A third and a weaker equivalence relation $T_1 \simeq T_2$ defined in [7] utilizes the extended real-valued seminorm introduced in the next section.

The integral above can be viewed as a $\bar{\mathcal{C}}$-valued bilinear form $<,>$ defined on $\mathcal{G} \times \mathcal{D}$:

$$<\psi, T>:= \int_\Omega (\psi \odot T)(x)dx = \int_\Omega f_T(\phi, x)\psi(x)dx + \mathcal{J}. \tag{1.9}$$

As stated in [3] , (1.9) allows one to visualize Colombeau g.f.'s as generalized distributions in the Schwartz sense. It is well known that if $T \in \mathcal{D}' \subset \mathcal{G}$, then $<\psi, T>$ is $\mathcal{C}$-valued and $<\psi, T>= T(\psi) \in \mathcal{C}$ for all $\psi \in \mathcal{D}(\Omega)$.

$\bar{\mathcal{C}}$ is dependent on n via the test functions used. However if it is constructed by defining $\mathcal{E}_M^\circ$ in $\mathcal{C}^{\bar{\mathcal{A}}_0(\mathcal{R})}$, where $\bar{\mathcal{A}}_0(\mathcal{R})$ is given by the class of index sets of D), then it becomes independent of dimension, (c.f. [2], p. 31). In this case the bilinear form of (1.9) has to be modified as

$$<\psi, T>= \int_\Omega f_T(I_n^1\phi, x)\psi(x)dx + \mathcal{J},\ T \in \mathcal{G}(\Omega),\ \ \Omega \subset \mathcal{R}^n,\ \phi \in \bar{\mathcal{A}}_0(\mathcal{R}).$$

As $C^\circ(\mathcal{R}^n) \subset \mathcal{D}'$, its elements can be imbedded into $\mathcal{G}(\Omega)$ via (1.6). But it is known that as a consequence of Schwartz's celebrated 'Impossibility Result', this is only an imbedding of vector spaces and not of algebras. In fact if f_1, f_2 are in C° and F_1, F_2 are corresponding g.f.'s in $\mathcal{G}$, we have $F_1 \odot F_2 \approx f_1.f_2$.

2. Topology and Duality

Biagioni and Egorov state that (c.f. [2] pp.45 and [10]) topologies in $\bar{\mathcal{C}}$ and $\mathcal{G}$ have not played a role in applications except for the well-posedness of a specific Cauchy problem. Topological properties have not drawn attention either in the existing literature on the stochastic analysis in Colombeau algebras except in the works of Çapar([5], [6]), Çapar–Aktuğlu ([7], [8], [9]) and Martias ([13]). This is partly due to the fact that Colombeau g.f.'s were regarded initially as tools of solving deterministic and stochastic non-linear PDEs and here the first concern was singling out the associated linear distributions. The construction of solutions to such PDEs as elements in $\mathcal{G}$ as well as the investigation of associated linear distributions can often be achieved on the representatives that belong to spaces of smooth functions. Thus no need would arise for reference to the overall topological properties of Colombeau algebras. Another reason might be the difficulty of defining natural topologies in these algebras which are easy to work with and which characterize well their intrinsic features, in particular their behaviour along delta sequences. Nevertheless the measurability considerations, the construction and extension of measures are intimately related to the global topological properties of Colombeau algebras. [7], [8] and [9] concentrate on topologies which were proposed by Biagioni but not further elaborated. The reason for the choice of Biagioni topologies is two-fold:

i) They are natural.
ii) Although they are not vector space topologies, they result in structures close to locally convex topological vector spaces (l.c.t.v.s.).

In fact they are induced by extended real-valued seminorms. Consequently a duality theory very similar (but having some pathological features) to the one in l.c.t.v.s.'s involved with the bilinear $\bar{\mathcal{C}}$-form of (1.9) can be constructed, which is summarized below.
Çapar and Aktuğlu use (1.9) to define random g.f.'s on $\mathcal{D}$ in the form $<\psi, T(\omega)>$ which in turn yields a theory similar in many respects to the theory of cylindrical measures of Gelfand–Vilenkin and Badrikian.

In $\bar{\mathcal{C}}$ a base of neighborhoods of the origin is introduced by the following sets:

$$Q_\mu := \{z \in \bar{\mathcal{C}} : \exists h \in \mathcal{E}_M^\circ, \bar{z} = h + \mathcal{J}, \text{ such that } |z(\phi)| < \mu, \forall \phi \in \mathcal{A}_0\} \tag{2.1}$$

or equivalently

$$\begin{aligned} Q_\mu := \quad & \{\bar{z} \in \mathcal{C} : \exists h \in \mathcal{E}_M^\circ, \bar{z} = h + \mathcal{J}, \exists N \geq 0 \text{ such that } \forall \phi \in \mathcal{A}_N(\mathcal{R}), \\ & \exists \eta > 0, |h(\phi_\epsilon)| < \mu \text{ if } 0 < \epsilon < \eta\}. \end{aligned} \tag{2.2}$$

These also induce an $\bar{\mathcal{R}}$-valued seminorm $\|.\|$:

$$\|\bar{z}\|^- := \begin{cases} \inf\{\mu > 0 : \bar{z} \in Q_\mu\} & \text{if } \exists \mu > 0, \bar{z} \in Q_\mu, \\ +\infty & \text{if } \forall \mu > 0, \bar{z} \notin Q_\mu, \end{cases} \tag{2.3}$$

(some of the seminorm axioms may be satisfied as $+\infty = +\infty$). We observe that $Q_{\frac{\mu}{2}} \subset \{\bar{z} : \|\bar{z}\|^- \mu\} \subset Q_\mu$, and the uniform topology of $\bar{\mathcal{C}}$ is generated by the filter in $\bar{\mathcal{C}} \times \bar{\mathcal{C}}$ spanned by the sets $\{(\bar{x}, \bar{y}) \in \bar{\mathcal{C}} \times \bar{\mathcal{C}} : \|\bar{x} - \bar{y}\| < \mu\}$. This topology of $\bar{\mathcal{C}}$ is non-Hausdorff and also not a linear vector space topology, since the base $\{Q_\mu : \mu > 0\}$ consists of non-absorbing sets. In it every Cauchy sequence converges, and the completeness follows from the fact that it behaves like a pseudo-metrizable space (c.f. [11]). An essential feature of $\bar{\mathcal{C}}$ is that it is a disconnected space under this topology. The finite and infinite components are given by:

$$\bar{\mathcal{C}}_f = \{\bar{z} \in \bar{\mathcal{C}} : \|\bar{z}\|^- < \infty\} \quad \text{and} \quad \bar{\mathcal{C}}_\infty = \bar{\mathcal{C}} \setminus \bar{\mathcal{C}}_f. \tag{2.4}$$

$\bar{\mathcal{C}}_f$ is an infinite dimensional non-Hausdorff l.c.t.v.s. generated by a single $\mathcal{R}$-valued seminorm $\|.\|^-|_{\bar{\mathcal{C}}_f}$. It is a non-metrizable and non-separable space of second category which is not locally compact, but barrelled. It induces on $\mathcal{C}$ its usual topology. Bounded sets in $\bar{\mathcal{C}}_\infty$ should be defined through centralizers. Similarly if $K \subset\subset \Omega, N \geq 0$ and $\mu > 0$, we set:

$$\begin{aligned} Q_{K,N,\mu} := \quad & \{T \in \mathcal{G}(\Omega) : \exists R \in \mathcal{E}_M(\Omega), T = R + \mathcal{N}, \text{ such that} \\ & \sup\{|\partial^\alpha R(\phi, x)| \; \alpha \in N^n, |\alpha| \leq N, \phi \in \bar{\mathcal{A}}_0(\mathcal{R}^n), x \in K\} < \mu\}. \end{aligned} \tag{2.5}$$

These sets form a base of neighborhoods of the origin. We also have:

$$\|T\|_{K,N} := \begin{cases} \inf\{\mu > 0 : T \in Q_{K,N,\mu}\} & \text{if } \exists \mu > 0, T \in Q_{K,N,\mu}, \\ +\infty & \text{if } \forall \mu > 0, T \notin Q_{K,N,\mu}. \end{cases} \tag{2.6}$$

The base of neighborhoods of the origin in $\mathcal{G}(\Omega)$ given by (2.5) defines a topology which also results from the uniform structure related to (2.6). This topology is again non-Hausdorff and not a vector space topology. It induces on $\mathcal{C}^\infty(\Omega)$ its own Frechét topology, and every Cauchy sequence in it converges.

$\bar{\mathcal{C}}$ can be made a separated space if one considers the quotient:

$$\bar{\mathcal{C}}_1 = \bar{\mathcal{C}}/\mathcal{J}_1, \; \mathcal{J}_1 = \{\bar{z} : \|\bar{z}\|^- = 0\} = (\|.\|^-)^{-1}\{0\}. \tag{2.7}$$

The finite component $\bar{\mathcal{C}}_{1,f}$ of $\bar{\mathcal{C}}_1$ is now a Banach algebra. But since $\mathcal{J}_1$ is an ideal only on $\bar{\mathcal{C}}_{1,f}$, $\bar{\mathcal{C}}_1$ will not be an algebra. Nevertheless the equivalence relation $\simeq$ related to $\mathcal{J}_1$, i.e.,

$$\bar{z}_1 \simeq \bar{z}_2 \Leftrightarrow \|\bar{z}_1 - \bar{z}_2\|^- = 0, \tag{2.8}$$

will be particularly useful in the sequel. It thereby defines another equivalence relation in $\mathcal{G}(\Omega)$:

$$T_1 \simeq T_2 \quad \Longleftrightarrow \quad \| < \psi, T_1 - T_2 > \|^- = 0, \; \forall \psi \in \mathcal{D}(\Omega). \tag{2.9}$$

We have $\sim \Rightarrow \approx \Rightarrow \simeq$ and also $\bar{z} = \mathcal{J} \Rightarrow \bar{z} \vdash 0 \Rightarrow \|\bar{z}\|^- = 0$.

Replacing the scalar field by $\bar{\mathcal{C}}$, a duality theory similar to that of one in t.v.s.'s can be established (c.f. [7], [8]). If E is a vector space, $\bar{E}^*$ (resp. $\bar{E}'$) will denote a $\mathcal{C}$-vector space of $\bar{\mathcal{C}}$-valued linear (resp. linear and continuous) functionals. It is also a $\bar{\mathcal{C}}$ (resp. $\bar{\mathcal{C}}_f$)-module. If E_n is an n-dimensional Euclidean space, then $\bar{E}_n^*$ and $\bar{E}_n'$ are not finite dimensional, in fact $\bar{E}_n'$ is a proper subspace of $\bar{E}_n^*$, e.g., consider $\bar{E}_1^*$ and $f(x) = x\delta(0)$. Then obviously $f \in \bar{E}_1^*$, but it can not be continuous since $\|\delta(0)\|^- = +\infty$. A $\bar{\mathcal{C}}$-dual pair (E, F) of vector spaces is defined in relation to a $\bar{\mathcal{C}}$-bilinear form $< .,. >$ with the usual separation axioms. For $(E, \bar{E}^*)$ and

$(E, \bar{E}')$, the separation axioms hold since a $\mathcal{C}$-form is also a $\bar{\mathcal{C}}$-form. Thus they are $\bar{\mathcal{C}}$-dual pairs. The weak topology $\bar{\sigma}(E,F)$ is the coarsest topology under which all $\mathcal{C}$-linear $\bar{\mathcal{C}}$-forms $f_y(.) =< ., y >, y \in F$ are continuous. The continuous $\bar{\mathcal{C}}$-dual of E under $\bar{\sigma}(E,F)$ is not necessarily equal to F. If we define $\bar{z}f$ $(f \in F)$ by $< x, \bar{z}f >= \bar{z} < x, f >$ and consider $\bar{z} \in \bar{\mathcal{C}}_f$, then $\bar{z}f$ is also continuous under the weak topology, but F may not be closed under this module operation. However if E is a t.v.s., then the $\bar{\mathcal{C}}$-dual of E under $\bar{\sigma}(E, E')$ is obviously $\bar{E}'$. In this case all $\bar{\mathcal{C}}$-forms take values in $\bar{\mathcal{C}}_f$ since a t.v.s. is always connected. The finite part of E denoted by E_f is defined as $E_f = \{x \in E :< x, y >\in \bar{\mathcal{C}}_f, \forall y \in F\} = \cap_{y \in F} f_y^{-1}(\bar{\mathcal{C}}_f)$.

We are more interested in $\mathcal{D} \longleftrightarrow \mathcal{G}$ duality. By (9) $\mathcal{G} \subset \bar{\mathcal{D}}^*$ via the bilinear form $< \psi, T >$, $\psi \in \mathcal{D}$, $T \in \mathcal{G}$. Nevertheless $(\mathcal{D}(\Omega), \mathcal{G}(\Omega))$ is not a $\bar{\mathcal{C}}$-dual pair. One half of the separation properties follows from $\mathcal{D}' \subset \mathcal{G}$. However $(\mathcal{D}, \mathcal{G}/\sim)$, $(\mathcal{D}, \mathcal{G}/\approx)$ and $(\mathcal{D}, \mathcal{G}/\simeq)$ are all $\bar{\mathcal{C}}$-dual pairs. Here, except in the first pair, the bilinear forms should be considered as $\bar{\mathcal{C}}_1$-valued.

In general $< \psi, T >$ is not continuous on $\mathcal{D}$ in its usual topology. However $\bar{\mathcal{D}}'$ is sufficiently rich and contains, besides the Schwartz distributions, many of the important generalized functions like $< \psi, (H^m \odot \delta)^{(k)} >$, ($(.)^{(k)}$ denotes differentiation in $\mathcal{G}$), c.f. [7].

As continuous functionals take values only in $\bar{\mathcal{C}}_f$ (or $\bar{\mathcal{C}}_{1,f}$) many parts of the duality theory in l.c.t.v.s. remain intact in $\bar{\mathcal{C}}$ duality with slight modifications. For instance if the annihilators and polars are defined using $\|.\|^-$, we have $\bar{E}'/\bar{M}^\perp \cong \bar{M}'$ algebraically and topologically (M being a finite dimensional subspace of E). Likewise if E is a separated l.c.t.v.s. and U is a neighborhood of the origin, then the $\bar{\mathcal{C}}$-polar $\bar{U}^\circ$ is $\bar{\sigma}(\bar{E}', E)$-compact etc.

3. Random Elements in $\bar{\mathcal{C}}$ and $\mathcal{G}(\Omega)$

Random Elements in $\bar{\mathcal{C}}$

Let (Δ, Σ, P) be a complete probability space. Random elements can be considered in spaces $\bar{\mathcal{C}}$, $\mathcal{G}(\Omega)$ and $\mathcal{G}(\mathcal{R}_+ \times \Omega)$. Different concepts of measurability have been considered by different authors. These concepts are mainly strong, weak, Borel and representative measurabilities. For mappings $Y : \Delta \to \bar{\mathcal{C}}$, we may have:

1) Strong measurability if there exists a sequence of $\bar{\mathcal{C}}$-valued simple functions $\{Y_m\}, (m = 1, 2, \cdots)$ such that $\lim_{m\to\infty} \|Y_m - Y\|^- = 0$ $P - a.s.$
2) Representative measurability if almost surely in $\omega \in \Delta$ we have a representation $Y(\omega) = z(\omega) + \mathcal{J}$ such that, for every $\phi \in \bar{\mathcal{A}}_0$, the complex-valued function $z(\omega, \phi)$ is measurable.
3) $\Sigma \longleftrightarrow \mathcal{B}_{\bar{\mathcal{C}}}$ measurability.
4) Weak measurability in the case $Y(\Delta) \subset \bar{\mathcal{C}}_{1,f}$. Then it is defined in the usual way using the elements of the $\mathcal{C}$-dual of the B-space $\bar{\mathcal{C}}_{1,f}$.

As usual 1) $\Longrightarrow$ 3) + separable range and whenever $Y(\Delta) \subset \bar{\mathcal{C}}_{1,f}$ we have 1) $\Longleftrightarrow$ 4) + separable range (c.f. [8], [9]).

The relation between 1) and 2) is especially important in stochastic problems. Unfortunately, in the utmost generality we do not have implication in either direction. This rather surprising conclusion stems from the fact that even a constant function $Y(\omega) = \bar{z}_0 \in \bar{\mathcal{C}}$ may possess non-constant (unfaithful) representatives, i.e., for some $\omega_1, \omega_2;\ \omega_1 \neq \omega_2$ we may have $z(\omega_1) \neq z(\omega_2)$ but $clz(\omega_1) = clz(\omega_2) = \bar{z}_0$. Thus $\bar{\mathcal{C}}$-valued simple functions may have representatives which are not scalar-valued simple functions. Under some restrictive assumptions the implications will be valid. For 1) $\Longrightarrow$ 2) we need the existence of faithful representatives of $Y_m(\omega)$ and a representative of $Y(\omega)$, say $z_m(\omega)$ and $z(\omega)$ satisfying $\lim_{m\to\infty} |z_m(\omega,\phi) - z(\omega,\phi)| = 0$ for all $\phi \in \bar{\mathcal{A}}_0$. For the other direction of the implication, uniformity in ϕ is required.

In accordance with 1) and 2) one can define the strong and representative integrabilities. The former is defined in Bochner fashion, but the limit $\int_\Delta Y dP(\omega) = \lim_{m\to\infty} \int_\Delta Y_m dP(\omega)$ should be considered in $\bar{\mathcal{C}}_1$, (i.e., within $\simeq$ equivalence).

The latter is defined through a Lebesgue integrable representative $z(\omega,\phi)$ such that $\int_\Delta z(\omega,\phi)dP(\omega)$ is moderate. Then:

$$E(Y) \simeq \int_\Delta z(\omega,\phi)dP(\omega) + \mathcal{J}, \tag{3.1}$$

and $E(Y)$ is well defined.

For the equivalence of two types of integrals, again the restrictive assumptions imposed for implications between measurability typea 1) and 2) are required. However in the following important special case the two concepts of integrability coincide:

Let $Y_{\psi,T}(\omega) =< \psi, T(\omega) >= [T(\omega)](\psi)$, where $T(\omega) : \Delta \to \mathcal{D}'(\mathcal{R}^n)$ is a weakly measurable mapping. Then under some natural assumptions:

$$\int_\Delta < \psi, T(\omega) > dP(\omega) \simeq E(Y_{\psi,T}) = g_\psi + \mathcal{J}_1 \in \bar{\mathcal{C}}_1.$$

If $T(\omega) = f_T(\omega) + \mathcal{N}$, then $g_\psi(\phi) = \int_\Delta \int_{\mathcal{R}^n} f_T(\omega,\phi,x)\psi(x)dxdP(\omega)$ (c.f. [8]).

Random Elements in $\mathcal{G}(\Omega)$

A Colombeau random generalized function (C.r.g.f.) is a measurable map $T : \Delta \to \mathcal{G}(\Omega)$. Again different types of measurabilities can be considered.

I) ([20]) T is measurable if there is a function $R_T : \Delta \times \mathcal{A}_1(\mathcal{R}^n) \times \Omega \to \mathcal{C}$ (or $\mathcal{R}$) with the following properties:
   i) For fixed $\phi \in \mathcal{A}_1(\mathcal{R}^n)$, $(x,\omega) \to R_T(\omega,\phi,x)$ is jointly measurable on $\Omega \times \Delta$.
   ii) Almost surely $\phi \to R_T(\omega,\phi,.)$ belongs to $\mathcal{E}_M(\Omega)$ and is a representative of $T(\omega)$. (Index sets in C) have been used.)

II) ([8]) $T : \Delta \to \mathcal{G}(\Omega)$ is weakly measurable if $\forall\psi \in \mathcal{D}(\Omega)$, $< \psi, T(\omega) >$ is representative (resp. strong) measurable. This generalizes the weak measurability of random Schwartz distributions in a natural way.

Denote all C.r.g.f. by $\mathcal{G}_\Delta(\Omega)$ and random Schwartz (Gelfand–Vilenkin) distributions by $\mathcal{D}'_\Delta(\Omega)$. In the two approaches the imbedding $\mathcal{D}'_\Delta(\Omega) \to \mathcal{G}_\Delta(\Omega)$ is achieved by:

I) If $T \in \mathcal{D}_\Delta(\Omega), \phi \in \mathcal{A}_1(\mathcal{R}^n)$, then $T(\omega) * \phi(x) = < T(\omega), \phi(x-.) >$ is measurable with respect to $\omega \in \Delta$ and smooth with respect to $x \in \mathcal{R}^n$, hence jointly measurable. Also $(T(\omega))_{\phi \in \mathcal{A}_1}$ belongs to $\mathcal{E}_M(\mathcal{R}^n)$. Thus $R_T(\omega, \phi, x) = T(\omega) * \phi(x)$ can be taken as a representing function of an element in $\mathcal{G}_\Delta(\Omega)$.

II) This is a straightforward imbedding, the measurability following from the weak measurability of random Schwartz distributions.

For $T \in \mathcal{G}_\Delta(\Omega)$, the integral $\int (T(\omega) dP(\omega)$ can be defined likewise, e.g., as a representative-type integral subject to the moderateness of

$$\int R_T(\omega, \phi, x) dP(\omega).$$

4. Generalized Stochastic Processes

These can be defined as classical processes with the usual index set $[0, \infty)$ or generalized index set $\mathcal{D}$. In both cases the process may be scalar, vector or $\mathcal{G}$-valued.

Oberguggenberger ([17]) attempts the construction of Colombeau algebras of stochastic processes (s.p.'s). These are differential algebras of generalized s.p.'s defined pathwise. Using the index functions given in A), the starting object in the deterministic case is $\mathcal{E}(\Omega) = (\mathcal{C}^\infty(\Omega))^N$. A sequence $\{f_\nu\}_{n \in N} \in \mathcal{E}(\Omega)$ is moderate if for all compact $K \subset\subset \Omega$ and all $\alpha \in N_0^n, \exists N > 0 \ni \sup_{x \in K} |\partial^\alpha f_\nu(x)| = O(\nu^N)$ as $\nu \to \infty$. Furthermore, f is said to be null if for all $K \subset\subset \Omega$, for all $\alpha \in N_0^n$ and all $N > 0$, $\sup_{x \in K} |\partial^\alpha f_\nu(x)| = O(\nu^{-N})$ as $\nu \to \infty$. (Here ν corresponds to $\phi \in \mathcal{A}_1$ or $\phi \in \bar{\mathcal{A}}_1$ of the previous constructions.) $\mathcal{G}(\Omega)$ is again of the form $\mathcal{G}(\Omega) = \mathcal{E}_M(\Omega)/\mathcal{N}(\Omega)$.

Any given $\phi \in S(\mathcal{R}^n)$ with $\int_{\mathcal{R}^n} \phi(x) dx = 1$ produces a mollifying sequence $\phi_\nu(x) = \nu^n \phi(\nu x)$ and an imbedding of compactly supported distributions. Thus in this case there is no privileged imbedding of $\mathcal{D}'$ into $\mathcal{G}$. The algebra of classical s.p.'s with a.s. continuous (resp. smooth) paths will be denoted by $\mathcal{CP}$ (resp. $\mathcal{C}^\infty\mathcal{P}$).

An element in $\mathcal{CP}$ (resp. $\mathcal{C}^\infty\mathcal{P}$) is a classical s.p., i.e., a map $\mathcal{R}^+ \longrightarrow \mathcal{L}^\circ(\Delta, \Sigma, P)$ with a.s. continuous (resp. smooth) paths. The basic objects are sequences of processes with a.s. smooth paths: $\mathcal{EP} = (\mathcal{C}^\infty\mathcal{P})^N$.
This is the space of all $X : \Delta \times N \times \mathcal{R}^+ \to \mathcal{R}$ such that

i) $\omega \to X_\nu(t, \omega)$ is measurable for all $\nu \in N$ and for all $t \in \mathcal{R}^+$,
ii) $t \to X_\nu(t, \omega)$ belongs to $\mathcal{C}^\infty(\mathcal{R}), \forall \nu \in N$ and almost all $\omega \in \Delta$.

$\mathcal{EP}$ is an algebra under component-wise multiplication and differentiation. We analogously have the subalgebra:

$$\mathcal{EP}_M = \{(X_\nu)_{\nu\in N} \in \mathcal{EP} : \text{ for almost all } \omega \in \Delta, \forall\tau > 0, \forall j \in N_0, \exists N(\omega),\\ C(\omega), \eta(\omega) > 0 \text{ with } \sup_{t\in[-\tau,\tau]} |\partial^j X_\nu(t,\omega)| \le C(\omega)\nu^{N(\omega)},\ \nu \ge \eta(\omega)\},$$

and the differential ideal $\mathcal{NP}$:

$$\mathcal{NP} = \{(X_\nu)_{\nu\in N} \in \mathcal{EP} : \text{ for almost all } \omega \in \Delta, \forall\tau > 0, j \in N_0, \forall N > 0,\\ \exists C(\omega), \eta(\omega) > 0 \text{ with } \sup_{[-\tau,\tau]} |\partial^j X_\nu(t,\omega)| \le C(\omega)\nu^{-N}, \nu \ge \eta(\omega)\}.$$

The differential algebra $\mathcal{GP}$ of generalized s.p.'s is

$$\mathcal{GP} = \mathcal{EP}_M/\mathcal{NP}.$$

Thus a generalized s.p. is an equivalence class of a three-parameter family $(\nu, t, \omega) \to X_\nu(t,\omega)$, where t and ω have their usual meanings and ν may be considered as the degree of singularity.

We can specify (unprivileged) imbeddings of the continuous path processes $\mathcal{CP}$ into $\mathcal{GP}$. Choose an element $\phi \in S(\mathcal{R})$ with $\int_{\mathcal{R}} \phi(s)ds = 1$ and all vanishing, i.e., $\int_{\mathcal{R}} s^j\phi(s)ds = 0$, $\forall j \ge 1$. Further we take $\psi \in \mathcal{D}(\mathcal{R})$ such that $\psi(s) \equiv 1$ in a neighborhood of $s = 0$. Given a process Z with a.s. continuous paths, we define the corresponding element of $\mathcal{GP}$ as

$$i(Z) = cl(Z * (\psi\phi_\nu))_{\nu\in N},$$

where $\phi_\nu(s) = \nu\phi(\nu s)$ and the convolution is pathwise. It easily follows that $(Z * \psi\phi_\nu)(t,\omega) = \int_{\mathcal{R}} Z(s,\omega)\psi(t-s)ds$ yields an element of $\mathcal{EP}_M$ so we can take its class in $\mathcal{GP}$. As an example, for a Brownian motion W we have $W \in \mathcal{CP}$, and $\xi = \frac{d}{dt}i(W)$ represents white noise as an element of $\mathcal{GP}$.

On the other hand the processes in $\mathcal{C}^\infty\mathcal{P}$ with almost surely smooth paths can be imbedded into $\mathcal{GP}$ by means of the standard imbedding $\sigma Z = cl((Z)_{\nu\in N})$ as well. Then the consistency property $i|_{\mathcal{C}^\infty\mathcal{P}} = \sigma$ holds (c.f. [17], Prop.3.4). In order to evaluate a generalized s.p. at fixed time t,we can introduce the concept of a generalized random variable (r.v.) in the same fashion as previous constructions, i.e., $\mathcal{ER} = (\mathcal{L}^\circ)^N$ being the space of sequences of measurable functions on Δ. We define the moderate and null sequences in a similar way to arrive at the algebra $\mathcal{GR} = \mathcal{ER}_M/\mathcal{NR}$. If $X \in \mathcal{GP}$ is a generalized s.p. and $t_0 \in \mathcal{R}$, then $X(t_0)$ is a generalized r.v. in $\mathcal{GR}$.

Oberguggenberger also outlines an approach which utilizes the topological-algebraic properties of $\mathcal{L}^\circ$. Thus rather than basing the construction on sample path properties, we can base it on sequences of smooth $\mathcal{L}^\circ$-valued functions on $\mathcal{R}$. In fact $\mathcal{L}^\circ$ equipped with convergence in probability is a complete, metrizable t.v.s. With respect to pointwise multiplication and superposition it is also a topological algebra and $\mathcal{C}^\infty(\mathcal{R},\mathcal{L}^\circ)$ is an algebra. Hence it makes sense to start with $\mathcal{E}(\mathcal{R},\mathcal{L}^\circ) = \{\mathcal{C}^\infty(\mathcal{R},\mathcal{L}^\circ)\}^N$ and end up again in a differential algebra of generalized $\mathcal{L}^\circ$-valued functions of the form:

$$\mathcal{G}(\mathcal{RL}^\circ) = \mathcal{E}_M(\mathcal{R},\mathcal{L}^\circ)/\mathcal{N}(\mathcal{R},\mathcal{L}^\circ).$$

The space $\mathcal{C}(\mathcal{R}, \mathcal{L}^2)$ of classical second order processes is imbedded into $\mathcal{E}(\mathcal{R}, \mathcal{L}^\circ)$ via

$$i : Z \longrightarrow (Z * \psi\phi_\nu)_{\nu \in N}$$

as before. The convolution integral here should be understood as an $\mathcal{L}^2$-valued Bochner integral. $i(Z)$ is of L^2-type, i.e., for each ν, $Z * \psi\phi_\nu \in \mathcal{C}(\mathcal{R}, \mathcal{L}^2)$. This imbedding also makes sense for $Z \in \mathcal{D}'(\mathcal{R}, \mathcal{L}^2)$. For $Z \in \mathcal{C}^\infty(\mathcal{R}, \mathcal{L}^2)$, the consistency result $(Z * (\psi\phi_\nu) - Z)_{\nu \in N} \in \mathcal{N}(\mathcal{R}, \mathcal{L}^\circ)$ holds.

Martias ([14]) attempts a straightforward generalization of Colombeau distributions to abstract Wiener spaces, i.e., an infinite dimensional version of Colombeau algebras and shows that this construction can be used for the representation of other well-known distributions on the Wiener space.

Let (H, B, μ) be an abstract Wiener space, H being the associated Cameron–Martin space. D will be the space of real-valued functions on B with bounded support and infinitely continuously $\mathcal{H}$ (Gross)- differentiable. E will denote the space of real-valued functions ϕ on B, infinitely $\mathcal{H}$-differentiable and satisfying the following condition: For any positive constant c, $\exists k$ and M such that $\| \nabla^n \phi \|_{L^n(H)} < kM^n$ for $\|x| \leq c$, where $L^\circ(H) = \mathcal{R}$, $L^n(H) = L(H, L^{n-1}(H))$, and ∇^n denotes the n-th order $\mathcal{H}$-derivative. Accordingly the counterpart of Colombeau's construction proceeds as follows: The index sets $\mathcal{A}_q$, $(q = 1, 2, \ldots)$ are

$$\mathcal{A}_q = \{\phi \in D : \int_B \phi d\mu = 1, \int_B (< y, x >)^k \phi(x)\mu(dx) = 0, y \in B', k = 1, 2, \ldots\}.$$

We then have $D \supset \mathcal{A}_1 \supset \mathcal{A}_2 \supset \cdots$.

To show that $\mathcal{A}_q \neq \emptyset$ for $q = 1, 2, \ldots$, Martias uses an integration by parts formula involved with $\mathcal{H}$-differentials. The families of E functions $E^{D \times \mathcal{R}^+}$ will replace $(\mathcal{C}^\infty)^{\mathcal{A}}$ in the original Colombeau construction. Its elements will be indicated by $f_{\phi,\epsilon}$. The algebra of moderate families denoted $E_M^{D \times \mathcal{R}^+}$ will consist of those f that satisfy: $\forall n \in N, \forall K$ (bounded) $\subset B, \exists p \in N, \exists \eta > 0$, such that $\forall \phi \in \mathcal{A}_p$, $\| \nabla^n f_{\phi,\epsilon}(x) \| < \epsilon^{-p}$, $x \in K$ for $0 < \epsilon < \eta$. A constant sequence $f_{\phi,\epsilon}(x) = g(x), x \in B, (\phi, \epsilon) \in D \times \mathcal{R}^+$ is obviously moderate. On the other hand the family $f_{\phi,\epsilon}(x) = \exp\{(\frac{1}{\epsilon})\phi(\frac{x}{\epsilon})\}$, $x \in \mathcal{R}$, $(\phi, \epsilon) \in D \times \mathcal{R}^+$ is in $E^{D \times \mathcal{R}^+}$, but it is not moderate.

Schwartz distributions (on Wiener space), Hida and Watanabe distributions are in the algebra of moderate families through regularizations. For defining Schwartz distributions on Wiener space, we may furnish D with the topology of uniform convergence on all bounded subsets of B along with H derivatives of all orders and consider the dual space D'. Thus the class of imbeddings into $E^{D \times \mathcal{R}^+}$ of Schwartz and Watanabe distributions will be representations of these distributions. Here again $G = E_M^{D \times \mathcal{R}^+} / E_0^{D \times \mathcal{R}^+}$, where the ideal $E_0^{D \times \mathcal{R}^+}$ consists of all families in $E^{D \times \mathcal{R}^+}$ with the property that $\forall K$ (bounded) $\subset B, \forall p \in N^+$, and $n \in N_0, \exists q \in N^+$ $(q = q(K, p))$ such that $\forall \phi \in \mathcal{A}_q$, we have $\| \nabla^n f_{\phi,\epsilon}(x) \| < \epsilon^p, x \in K$, this being valid for all $\epsilon \in (0, \epsilon_0)$ for some $\epsilon_0 > 0$.

Martias in his two other papers ([13], [15]) studies the following types of stochastic integrals to be used subsequently in the generalized counterparts of the Ito calculus and stochastic PDE:

$$(1)\ \int_0^t T(X_s)dW_s, \qquad (2)\int_0^t T_s dW_s.$$

In (1) $T \in \mathcal{G}(\Omega)$ is a Colombeau g.f. and X is any continuous s.p. taking values in the open set $\Omega \subset \mathcal{R}^n$. The resulting process is a continuous martingale with values in $\bar{\mathcal{C}}$. The integral is constructed representative-wise. Let $f_T(\phi, x)$ be any representative of T. Then $\int_0^t f_T(\phi, X_s)dW_s$ is well defined as an ordinary stochastic integral with respect to W. Under the assumptions the integral yields definitely a moderate element in E_M°. To show that the integral is independent of the choice f_T of the representative, one has to verify that $g \in \mathcal{N}(\Omega)$ yields $\int_0^t g(\phi, X_s)dW_s \in \mathcal{J}$ *a.s.* However due to the erroneous proof of Lemma 3.1.1. of the paper [13] , one can only show that $\int_0^t T(X_s)dW_s$ is uniquely determined in $\bar{\mathcal{C}}_1$.

A similar result applies to the construction of the stochastic integral (2). Here $(T_t)_{t\geq 0}$ will be a process with values in $\mathcal{G}(\Omega)$. It is supposed that this process is well-measurable in the sense of the general theory of s.p.'s taking values in a topological space and moreover locally square-integrable (for $\int_\Delta T_t^2 dP$, c.f. section 3). Using any ordinary process $f_t^T(\phi, x), \phi \in \mathcal{A}$ (in the index set system B) of the Introduction) such that $T_t = cl[f_t^T(\phi, x)]$, we define $\int_0^t T_s dB_s$ as the class (in $\bar{\mathcal{C}}_1$) of $\int_0^t f_s^T(\omega, \phi, x)dW_s(\omega)$.

Unlike Schwartz distributions, Colombeau distributions have point values. If $T \in \mathcal{G}(\Omega)$ has a representation $T = R_T + \mathcal{N}$, then $T(x_0)$ is defined as the generalized constant $T(x_0) = R_T(\phi, x_0) + \mathcal{J} \in \bar{\mathcal{C}}$. Taking this point into account, Martias ([15]) goes through a straightforward generalization of the following classical form of the Ito formula for a real-valued function $f \in \mathcal{C}^2$:

$$f(W_t) = f(0) + \int_0^t \nabla f(W_s)dW_s + \frac{1}{2}\int_0^t \triangle f(W_s ds. \tag{4.1}$$

(4.1) had been extended before by Üstünel ([23]) to the tempered distributions via the stochastic calculus of variation.

Indicating equivalence classes in $\mathcal{G}$ by [ ], for $T = [f_T]$, (4.1) implies:

$$[f_T(\phi, W_t)] = [f_T(\phi, 0)] + \Big[\int_0^t \nabla_x f_T(\phi, W_s)dW_s\Big] + \frac{1}{2}\Big[\int_0^t \triangle f_T(\phi, W_s)ds\Big].$$

The moderateness of integrals on the right-hand side follows from easy estimates. If one poses

$$\int_0^t [\nabla_x f_T(\phi, W_s)]dW_s = \Big[\int_0^t \nabla_x f_T(\phi, W_s)dW_s\Big],$$

$$\int_0^t [\triangle f_T(\phi, W_s)]ds = \Big[\int_0^t \triangle f_T(\phi, W_s)ds\Big],$$

the following equality in $\bar{\mathcal{C}}$ is obtained:

$$[f_T(\phi, W_t)] = [f_T(\phi, 0)] + \int [\nabla_x f_T(\phi, W_s)] dW_s + \frac{1}{2}\int_0^t [\Delta f_T(\phi, W_s)] dW_s$$

which can be rewritten as:

$$T(W_t) = T(0) + \int_0^t \nabla T(W_s) dW_s + \frac{1}{2}\int_0^t \Delta T(W_s) ds, \tag{4.2}$$

and which is the generalization considered. In particular if $G \in \mathcal{D}'(\Omega)$, then substituting $T = i(G)$, (21) generalizes the Ito formula to all Schwartz distributions.

Çapar and Aktuğlu ([7], [8]) define a Colombeau random g.f. (C.r.g.f.) as a weakly measurable map $T : \Delta \to \mathcal{G}(\Omega)$ (c.f. Section 3-II). Weak measurability is in the sense of $(\mathcal{G}, \mathcal{D})$ pseudo-duality, which means $\forall \psi \in \mathcal{D}(\Omega)$, $< \psi, T(\omega) >$ is measurable (usually representative-wise), i.e., if there exists a representative $f_T(\omega, \phi, x)$ of $T(\omega)$ such that $h_T(\omega, \phi) = \int_{\mathcal{R}^n} f_T(\omega, I_n^1 \phi, x)\psi(x)dx$ is measurable for every $\phi \in \bar{A}_0(\mathcal{R})$. (See (7) for I_n^1.) If $T \in \mathcal{D}'_\Delta(\mathcal{R}^n)$, we have $< \psi, T(\omega) > = [T(\omega)](\psi) \in \mathcal{C}$, and the ordinary measurability of $< \psi, T(\omega) >$ coincides with the representative measurability. Hence we recover the weak measurability of random Schwartz distributions.

A generalized stochastic process (g.s.p.) in their sense is a map $X : \mathcal{D}(\Omega) \to \bar{\mathcal{L}}(\Delta, \Sigma, P)$ which is linear when considered as a mapping into $\bar{L}(\Delta, \Sigma, P)$. (Here $\bar{L}$ is the $\mathcal{C}$-linear space of measurable $\bar{\mathcal{C}}$-valued functions (also a $\bar{\mathcal{C}}$-module) and $\bar{L}$ denotes its equivalence classes under $P - a.s.$ equality). Of course for $T \to \mathcal{G}(\Omega)$ as above, $< \psi, T(\omega) >$ is continuous (a.s. or path-wise), then it takes values in $\bar{\mathcal{C}}_f$, thus the theory resembles that of the classical linear random distributions.

They first try to find the support of the process distribution. For this purpose the restriction of the g.s.p. to finite dimensional subspaces is considered. If $\mathcal{F} \subset \mathcal{D}$ is such a finite dimensional subspace and X is a.s. continuous, let $X|_{\mathcal{F}} : \mathcal{F} \to \bar{\mathcal{L}}(\Delta, \Sigma, P)$ be its restriction to $\mathcal{F}$. Then it induces a probability measure $\nu_{\mathcal{F}}$ in $\bar{\mathcal{F}}' \cong \bar{\mathcal{F}}_f^* \cong \bar{\mathcal{D}}'/\bar{\mathcal{F}}^\perp$. Here $\bar{\mathcal{F}}_f^*$ denotes all linear $\bar{\mathcal{C}}$ forms taking values in $\bar{\mathcal{C}}_f$ and it is now isomorphic to $\bar{\mathcal{F}}'$, since a linear mapping of a finite dimensional space into a t.v.s. is continuous (recall that $\bar{\mathcal{C}}$ is a t.v.s.). Let χ denote the family of all finite dimensional subspaces of $\mathcal{D}$. The following is proved in analogy to the classical case (c.f. [8]):

- The projective limit, $\lim_{\leftarrow} \bar{\mathcal{F}}'$ of the projective system $\{\bar{\mathcal{F}}\}_{\mathcal{F}} \in \chi$ under the canonical restrictions, is isomorphic to the finite part $\bar{\mathcal{D}}_f^*$ of the algebraic $\bar{\mathcal{C}}_1$-dual of $\mathcal{D}$.
- The inductive limit, $\lim_{\rightarrow} \mathcal{F}$ of all finite dimensional subspaces of $\mathcal{D}$, does not yield the standard inductive limit topology in $\mathcal{D}$, but rather $\mathcal{D}$ in its finest locally convex topology. Its continuous $\bar{\mathcal{C}}_1$-dual in this topology is $\bar{\mathcal{D}}^*$.
- The projective limit topology in $\bar{\mathcal{D}}_f^*$ is identical with $\bar{\sigma}(\bar{\mathcal{D}}_f^*, \mathcal{D})$.

- Every probability measure on the Borel σ-algebra of $\bar{\mathcal{F}}'$ is Radon. This is because, since $\bar{\mathcal{F}}'$ is metrizable, every probability measure on it is Carathéodory regular, i.e., regular over closed sets. As the weak and strong topologies coincide on $\bar{\mathcal{F}}'$, it is also Carathéodory regular in the weak topology of $\bar{\mathcal{F}}'$. As pointed out before, the unit ball as well as balls around the origin are $\sigma(\bar{\mathcal{F}}', \mathcal{F})$ -compact. Hence $\bar{\mathcal{F}}'$ is σ-compact in $\bar{\sigma}(\bar{\mathcal{F}}', \mathcal{F})$. Finally the weakly closed sets approximating a measurable set from inside can be intersected by expanding balls, resulting in an approximating sequence of weakly compact sets.
- We also note that the projective system $\{\bar{\mathcal{F}}'\}_{\mathcal{F}} \in \chi$ possesses the sequential maximality (s.m.) property.

Hence we arrive at the following result (Thm. 4.8 , [8]) which can be proved in the same fashion as the Bochner–Badrikian theorem in the classical theory of cylindrical measures. In fact the proof capitalizes upon only the Radon and s.m. property and the well-known results of the projective limits:

Let $\{\nu_{\mathcal{F}}\}_{\mathcal{F}\in\chi}$ be the projective system of probability measures on the Borel σ-algebra of $\bar{\mathcal{F}}'$ resulting from an a.s. continuous g.s.p. Then there exists a unique extension ν to $\bar{\mathcal{F}}' \cong \bar{\mathcal{D}}^*_f$ equipped with the smallest σ-algebra $\mathcal{H}$ making all canonical projections $\pi_{\mathcal{F}}$ of $\bar{\mathcal{D}}^*_f$ onto $\bar{\mathcal{F}}'$ measurable such that $\nu_{\mathcal{F}} = \pi_{\mathcal{F}}(\nu)$, $\forall \mathcal{F} \in \chi$.

If $\mathcal{G}_f(\Omega)$ denotes, in accordance with the definition in Section 2, all $T \in \mathcal{G}(\Omega)$ such that $< \psi, T > \in \bar{\mathcal{C}}_f$, $\forall \psi \in \mathcal{D}(\Omega)$, then $\mathcal{G}_f(\Omega) \subset \bar{\mathcal{D}}^*_f(\Omega)$ and the inclusion is dense. We clearly have $\mathcal{G}_f|_{\mathcal{F}} \cong \bar{\mathcal{F}}'$. If $p_{\mathcal{F}}$ denotes the restriction of $\pi_{\mathcal{F}}$ to $\mathcal{G}_f$, then the same family $\{\nu_{\mathcal{F}}\}_{\mathcal{F}\in\chi}$ can be used to create a cylindrical measure $p^{-1}_{\mathcal{F}}(\mathcal{F})$. As a result when the process X is a.s. path continuous, the ν-outer measure of $\mathcal{G}_f(\Omega)$ is 1, thus the probability distribution of the g.s.p. is supported by the trace of $\mathcal{H}$ in $\bar{\mathcal{D}}^*_f$ over $\mathcal{G}_f(\Omega)$. Thus as it is seen, one of the main advantages of modeling a g.s.p. as $< \psi, T(\omega) >$ is that it allows a natural extension of the Gelfand–Vilenkin theory of linear random distributions.

Another advantage of this approach is the ease it provides for treating white noise and its non-linear functions. For that purpose we use a canonical probability space (Δ, Σ, ν) where $\Delta = S'(\mathcal{R}^n)$ is the space of tempered Schwartz distributions on $\mathcal{R}^n$, Σ : the Borel σ-algebra in S' generated by the weak* topology and ν is the unique probability measure on S' provided by the Bochner–Minlos theorem, i.e., is the one that satisfies:

$$\int_{S'} e^{i<\phi,\omega>} d\nu(\omega) = e^{-\frac{1}{2}\|\phi\|_{L^2}}, \phi \in S(\mathcal{R}^n). \tag{4.3}$$

Then the white noise is the linear generalized process ξ, given by

$$\xi : \Delta \to \mathcal{D}'(\mathcal{R}^n), < \psi, \xi(\omega) > = < (-1)^n \frac{\partial^2 \psi}{\partial x_1, \cdots, \partial x_n}, W >$$

where W is an n parameter Wiener process, $\omega \in \Delta$ and $\psi \in \mathcal{D}(\mathcal{R}^n)$. Viewed as a g.s.p. in the Colombeau sense, it is of the form $< \psi, T_\xi(\omega) >$, where C.r.g.f. T_ξ

has a representative

$$f_\xi(\omega,\phi,x) =< \phi(.-x),\omega >, \phi \in \bar{\mathcal{A}}_0(\mathcal{R}^n).$$

The distribution of the process is supported by $S'(\mathcal{R}^n)$ and we note that

$$S'(\mathcal{R}^n) \subset \mathcal{D}'(\mathcal{R}^n) \subset \mathcal{G}_f(\mathcal{R}^n) \subset \bar{\mathcal{D}}_f^*.$$

As an illustration let $W(t), t \geq 0$ be a one-parameter Brownian motion. Suppressing ω, we have $\xi(\psi) =< \psi, T_\xi >= g_\psi + \mathcal{J}$ where

$$\begin{aligned} g_\psi(\phi) &= \int_{\mathcal{R}} f_\xi(\phi,x)\psi(x)dx \\ &= -\int_{\mathcal{R}^2} W(t)\phi'(t-x)\psi(x)dtdx,\ \phi \in \bar{\mathcal{A}}_0, \psi \in \mathcal{D}(\mathcal{R}^+). \end{aligned}$$

The $\bar{\mathcal{C}}_1$-valued mean functional $\bar{M}_\xi(\psi) := E[< \psi, T_\xi >] = m_{\xi,\psi} + \mathcal{J}_1$ and the covariance functional $\bar{K}_\xi(\psi_1,\psi_2) = E[< \psi_1, T_\xi >< \psi_2, T_\xi >] = k_{\xi;\psi_1,\psi_2} + \mathcal{J}_1$ can easily be calculated in the sense of representative integrals (c.f. (19)), here the two concepts of integrability coincide):

$$m_{\xi;\psi}(\phi) = E\{-\int_{\mathcal{R}^2} W(t)\phi'(t-x)\psi(x)dx\} = 0.$$

$$k_{\xi;\psi_1\psi_2}(\phi) = E\{\int_{\mathcal{R}^4} W(t_1)W(t_2)\phi'(t_1-x)\phi'(t_2-y)\psi_1(x)\psi_2(y)dt_1dt_2dxdy\}$$

$$= \int_{\mathcal{R}^4} \min(t_1,t_2)\phi'(t_1-x)\phi'(t_2-y)\psi_1(x)\psi_2(y)dt_1dt_2dxdy,\ \ \psi_1,\psi_2 \in h\mathcal{D}(\mathcal{R}^+).$$

This is not different from $E[\xi(\psi_1)\xi(\psi_2)] = \int_{\mathcal{R}} \psi_1(t)\psi_2(t)dt$ given by the linear distribution theory. In fact it can easily be shown that $k_{\xi;\psi_1,\psi_2} - \int_{\mathcal{R}} \psi_1(t)\psi_2(t)dt \in \mathcal{J}_1$.

Now the formal expression $\xi^2 = (\frac{dW}{dt})^2$ is a singular object which is impossible to depict in the linear Gelfand–Vilenkin theory, but straightforward with the above formulation:

$$\xi^2(\psi) =< \psi, T_{\xi^2} >= h_\psi + \mathcal{J},\ T_{\xi^2} = f_{\xi^2} + \mathcal{N} = (f_\xi)^2 + \mathcal{N},$$

where

$$f_{\xi^2}(\phi,x) = [\int_{\mathcal{R}} W(t)\phi'(t-x)dt]^2 = \int_{\mathcal{R}^2} W(t_1)W(t_2)\phi'(t_1-x)\phi'(t_2-x)dt_1dt_2,$$

and

$$h_\psi(\phi) = \int_{\mathcal{R}^3} W(t_1W(t_2)\phi'(t_1-x)\phi'(t_2-x)\psi(x)dt_1dt_2dx,$$

hence

$$M_{\xi^2}(\psi) = m_{\xi^2;\psi} + \mathcal{J}_1,$$

$$m_{\xi^2;\psi}(\phi) = \int_{\mathcal{R}^3} \min(t_1,t_2)\phi'(t_1-x)\phi'(t_2-x)\psi(x)dt_1dt_2dx,$$

$$\bar{K}_{\xi^2}(\psi_1,\psi_2) = k_{\xi^2;\psi_1,\psi_2} + \mathcal{J}_1,$$

where

$$k_{\xi^2;\psi_1,\psi_2}(\phi) = \int_{\mathcal{R}^6} E[\prod_{i=1}^4 W(t_i)]\phi'(t_1 - x)\phi'(t_2 - x)\phi'(t_3 - y)\phi'(t_4 - y)$$

$$\times\psi_1(x)\psi_2(y)\prod_{i=1}^4 dt_i dx dy,$$

and an easy calculation yields $E[\prod_{i=1}^4 W(t_i)] = t_{(1)}(2t_{(2)} + t_{(3)})$, $t_{(i)}$ denoting the ordered t_i. Likewise for any function $F \in \mathcal{C}^\infty(\mathcal{R})$, slowly increasing at infinity, the function $F(\xi)$ of the white noise is well defined as a C.r.g.f. In many applications we prefer white noise being defined as a genuine process. This can be done by considering

$$\xi_t(\psi) = \int_0^t \psi(s)dW(s) = -\int_0^t W(s)d\psi(s),$$

which is a usual process with values in $\mathcal{D}'$. Now the Colombeau version of this is given by $T_{\xi_t} = f_{\xi_t} + \mathcal{N}$, where

$$f_{\xi_t}(\omega, \phi, x) = -\xi_{t,s}(\omega)(\phi(s - x)) = -\int_0^t W(\omega, s)\phi'(s - x)ds,\ \phi \in \bar{\mathcal{A}}_0(\mathcal{R}).$$

5. Applications to Nonlinear Stochastic PDE

Colombeau's theory offers great flexibility and potentiality in dealing with nonlinear stochastic PDEs, especially for those having irregular, distributional or even generalized distributional initial data. First considering the deterministic case, let $L(D)U(x) = 0, x \in \Omega \subset \mathcal{R}^n$ be an mth order non-linear PDE, e.g., $U_t + U_x U = 0$ (first order shock-wave equation) or $\Box U = F(U)$ (non-linear wave equation, F is $\mathcal{C}^\infty$ or $\mathcal{C}^\infty$ and globally Lipshitz). If there is a $\mathcal{C}^\infty$ solution U, then $L(D)U$ as computed in $\mathcal{C}^\infty$ and in $\mathcal{G}(\Omega)$ will be identical. Thus we have $L(D)U = 0$ both in $\mathcal{C}^\infty(\Omega)$ and $\mathcal{G}(\Omega)$. On the other hand, if we have classical solutions $U \in \mathcal{C}^m(\Omega) \setminus \mathcal{C}^\infty(\Omega)$, then we still have $L(D)U = 0$ in $\mathcal{C}^\circ$, however $L(D)U$ computed in $\mathcal{G}(\Omega)$ and $\mathcal{C}^\circ(\Omega)$ will be no longer identical, but rather associated by the equivalence relation $\approx$. Thus, $L(D)U = 0$ in $\mathcal{C}^\circ(\Omega) \iff L(D)U \approx 0$ in $\mathcal{G}(\Omega)$. That is, $U \in \mathcal{C}^m(\Omega) \setminus \mathcal{C}^\infty(\Omega)$ is a classical solution $\iff$ it is a solution of $L(D)U \approx 0$. As a third case there may be non-classical generalized solutions $U \in \mathcal{D}'(\Omega)$ or $U \in \mathcal{G}(\Omega) \setminus \mathcal{D}'(\Omega)$. We must then consider $L(D)U \approx 0$ or even $L(D)U \sim 0$ (c.f. [21]). If $T \in \mathcal{D}'(\Omega)$ exists as a solution, the natural coherence relations $U \approx T$ and $U \sim T$ should be sought. That will also be the case in stochastic PDEs, e.g., when we consider solutions as stochastic processes with path space in $\mathcal{G}(\Omega)$, then we should investigate pathwise limits to find the associated linear distributions if there are any. The existence of such associated linear distributions would indicate the information contained in the generalized solution at the level of the linear theory.

To illustrate these ideas, we take up the equation discussed in [20], that is the stochastic semi-linear wave equation given by

$$\begin{aligned} \Box U &= F(U) + H \quad \text{on} \quad \mathcal{R}^{n+1}, \\ U|_{\{t<0\}} &= 0, \end{aligned} \tag{5.1}$$

where F is a smooth and globally Lipschitz function, $\Box$ denotes the d'Alembertian $\partial_t^2 - \partial^2 x_1 - \cdots - \partial^2 x_n$, and H is a Gelfand–Vilenkin random generalized function, i.e., a weakly measurable map $H : \Delta \to \mathcal{D}'(\mathcal{R}^{n+1})$ which vanishes on $\{t < 0\}$ a.s., ((Δ, Σ, μ) is a fixed probability space). It is known that the solution to the linear wave equation

$$\begin{aligned} \Box V &= H \quad \text{on} \quad \mathcal{R}^{n+1}, \\ V|\{t < 0\} &= 0 \end{aligned} \tag{5.2}$$

is a linear generalized process in many cases, for example when $n \geq 2$ and H is a space-time white noise on $E = \mathcal{R}^n \times [0, \infty)$. The authors show that given any C.r.g.f. H, the non-linear problem (5.1) has a unique Colombeau solution $U : \Delta \to \mathcal{G}(\mathcal{R}^{n+1})$ in space dimensions $n = 1, 2, 3$. The technique is to solve the equation at the representative platform and then show that the solution is a moderate function, thus its class in $\mathcal{G}(\Omega)$ is the required solution. For that purpose it is further assumed that $F(0) = 0$ and $F \in \mathcal{O}_M(\mathcal{R})$, i.e., a function slowly increasing at infinity. For a fixed $\eta > 0$, one can choose a representative $R_H^\eta(\omega, \phi, x, t)$ $(x \in \mathcal{R}^n)$ of H, such that if $supp H \subset E$, then $R_H^\eta \equiv 0$ when $t \leq -\eta$ for all $\phi \in \mathcal{A}(\mathcal{R}^{n+1})$ and almost all $\omega \in \Delta$, (the index system as in D)). Then we solve the classical PDE

$$\Box u(x, t) = F(u(x, t)) + R_H^\eta(\omega, \phi, x, t),$$

$$u(x, t) = 0 \quad \text{for} \quad t \leq -\eta\,, x \in \mathcal{R}^n.$$

Call the classical smooth solution $R_U^\eta(\omega, \phi, x, t)$. This solution is obtained by Picard iteration in the corresponding integral equation using the Lipschitz property of F and can be given explicitly e.g., in space dimension $n = 3$, by the Kirchhoff formulae and in dimensions $n = 1, 2$ by d'Alembert and Poisson formulae respectively. $R_U(\omega, \phi, x, t)$ is jointly measurable in (ω, x, t). The zeroth and higher order derivative estimates are obtained by using Gronwall's inequality showing that R_U^η is moderate. Its class $U(\omega)$ in $\mathcal{G}_\Delta(\mathcal{R}^{n+1})$ will define a solution to

$$\Box U = F(U) + H,$$

$$supp U \subset \mathcal{R}^n \times [-\eta, \infty).$$

It is in fact shown that $U(\omega)$ has its support contained in $\mathcal{R}^n \times [0, \infty)$ by demonstrating that the class of $R_U^\eta(\omega, \infty)$ does not depend on the choice of η, i.e., for two η's we have $R_U^{\eta_1}(\omega, \phi) - R_U^{\eta_2} \in \mathcal{N}(\mathcal{R}^{n+1})$. The uniqueness follows from the standard arguments.

In the case of white noise excitation, i.e., $H = \dot{W} = \xi$, the authors work out the pathwise limits explicitly. Now because of the extra dimension on the positive time axis we should consider $S(E) = S(\mathcal{R}^{n+1})|_E$, $\Delta = S'(E)$ and $T_\xi : \Delta \to \mathcal{D}'(\mathcal{R}^{n+1})$, with the representative $R_\xi(\omega, \phi, x, t) =< \phi(. - x, . - t|_E, \omega >, \phi \in$

proved that $(\frac{\partial^2}{\partial t^2} + F'(|X|^2)X = \dot{W}$ has a unique solution X which is a random variable taking values in $\mathcal{G}(\mathcal{R}^4)$. The proof utilizes two technical lemmas and an intermediate result regarding the unique solution of the same problem with smooth initial data.

In [17], Obberguggenberger deals with the application of his version of stochastic Colombeau algebras to a certain first order non-linear stochastic differential equation. To get around the difficulties inherent in measurability considerations, he uses the construction $\mathcal{G}(\Omega) = \mathcal{E}_M(\Omega)/\mathcal{N}(\Omega)$, $\mathcal{E}(\Omega) = (\mathcal{C}^\infty(\Omega))^N$ as discussed in the previous section. The equation he deals with is

$$\begin{aligned} \dot{X}(t) &= F(X(t), Y(t)), \ \ t \in \mathcal{R}, \\ X(0) &= A. \end{aligned} \tag{5.5}$$

The solution sought will be visualized as $X \in \mathcal{G}P$ (a generalized process in the Oberguggenberger sense):

$$X = Cl\{(X_\nu)_{\nu \in N}\}.$$

The driving term is also an element of $\mathcal{G}P$:

$$Y = Cl\{(Y_\nu)_{\nu \in N}\}.$$

The initial value is a generalized random variable

$$A = Cl\{(A_\nu)_{\nu \in N}\} \in \mathcal{G}R.$$

F may be given by a tempered sequence

$$F = \{(F_\nu)_{\nu \in N} \in \mathcal{E}(\mathcal{R}^2),$$

i.e., $F_\nu \in \mathcal{O}_M(\mathcal{R}^2)$ and $\forall \alpha \in N_0^2, \exists m, C > 0$ such that $|\partial^\alpha F_\nu(x,y)| \le \nu^m(1 + |x| + |y|)^m$, for all $(x,y) \in \mathcal{R}^2, \nu \in N$. The equation itself is to be understood as

$$(\dot{X}_\nu - F_\nu(X_\nu, Y_\nu))_{\nu \in N} \in \mathcal{N}P \ \text{ and } \ (X_\nu(0) - A_\nu)_{\nu \in N} \in \mathcal{N}R.$$

If this holds for one representative, it holds for all. Then the author proves that (5.5) has a unique solution under the assumption of $\{F_\nu\}$ being tempered and also having globally bounded gradient with logarithmic estimates. The method is again finding path-wise classical solutions. The classical theory then gives for every ν a unique solution X_ν with smooth paths, thus $\{X_\nu\}_{\nu \in N} \in \mathcal{E}P$. To prove the existence of a generalized solution, it suffices to show that $\{X_\nu\}_{\nu \in N} \in \mathcal{E}\mathcal{P}_M$. Then its class in $\mathcal{G}P$ will yield a solution. For this purpose using the MVT, we can write

$$\begin{aligned} X_\nu(t) = A_\nu + tF_\nu(0,0) &+ \int_0^t \partial_y F_\nu(\Theta(X_\nu(s), Y_\nu(s)))Y_\nu(s)ds \\ &+ \int_0^t \partial_x F_\nu(\Theta(X_\nu(s), Y_\nu(s)))X_\nu(s)ds \end{aligned}$$

for some function Θ. We note that $F_\nu(0,0)$ and $Y_\nu|_{[-T,T]}$ are $O(\nu^m)$ on any fixed time interval $[-T,T]$ and $\partial_x F_\nu$, $\partial_y F_\nu$ are $O(\log \nu)$. Then Gronwall's inequality gives

$$\sup_{t\in[-T,T]} |X_\nu(t) \leq C\nu^m e^{T\log\nu} = C\nu^{m+T}.$$

This completes the zero order estimate for $\{X_\nu\}$. Similar estimates ensue for the derivatives of X_ν of any order. For the uniqueness: easy inductive estimates yield $\{X_\nu - \tilde{X}_\nu\}_{\nu\in N} \in \mathcal{N}P$, thus $X = \tilde{X}$ in $\mathcal{G}P$. On the consistency part: in case the tempered generalized function F is given by $F_\nu \equiv F$ as a classical globally Lipschitz continuous function, the driving process S is smooth, i.e., $S \in \mathcal{C}^\infty\mathcal{P}$ and $A_\nu \equiv A \in \mathcal{L}^0$ is a classical r.v., then the problem

$$\dot{Z} = F(Z, S), \quad Z(0) = A$$

has a classical smooth solution Z and $X \in \mathcal{G}P$ satisfies $X = i(Z)$.

As an example, the author considers the multiplicative noise equation

$$\dot{X}(t) = \alpha X(t)\dot{W}(t)\ , \ X(0) = A \in \mathcal{G}R, \ \ (\dot{W} = \frac{d}{dt}(i(W)).$$

Although the r.h.s. here corresponding to F does not satisfy the hypotheses of the above result, there is still a unique solution $X \in \mathcal{G}P$ due to the local boundedness of the Brownian motion. The representatives yield the classical equation

$$\dot{X}_\nu(t) = \alpha X_\nu(t)\dot{W}_\nu(t) + J_\nu(t) \quad \text{with} \quad X_\nu(0) = A_\nu + I_\nu,$$

for some $J \in \mathcal{N}P$ and $I \in \mathcal{N}R$, where $\{A_\nu\}_{\nu\in N}$ represents A. This yields

$$X_\nu(t) = (A_\nu + I_\nu)\exp(\alpha W_\nu(t) - \alpha W_\nu(0)) + \int_0^t \exp(\alpha W_\nu(t) - \alpha W_\nu(s))J_\nu(s)ds.$$

Due to the pathwise boundedness of $W_\nu(t)$ on any compact interval independently of $\nu \in N$, the sequence $\{X_\nu\}_{\nu\in N}$ belongs to $\mathcal{E}\mathcal{P}_M$ and defines a solution. One representative of the solution can be taken as $\tilde{X}_\nu(t) = A_\nu \exp(\alpha W_\nu(t) - \alpha W_\nu(0))$. In case $A_\nu \equiv A$ is a classical r.v., $X_\nu(t)$ or $\tilde{X}_\nu(t)$ converges pathwise uniformly on $[-T, T]$ as $\nu \to \infty$ to $Z(t) = Ae^{\alpha W(t)}$, so that the generalized solution $X \in \mathcal{G}P$ is associated with the classical process $Z \in \mathcal{C}P$. In fact Z is the Stratanovich solution of the above multiplicative white noise problem.

Martias considers in [15]

$$X_t = \eta + \int_0^t G(s, X_s)dW_s + \int_0^t F(s, X_s)ds \tag{5.6}$$

as a straightforward generalization of the Ito differential equation to generalized-constant valued processes. Here $G, F : \mathcal{R}^+ \times \Delta \times \bar{\mathcal{C}} \to \bar{\mathcal{C}}$ and η is a $\bar{\mathcal{C}}$-valued r.v. The equality is to be understood in the sense of association $\vdash$ with the usual complex numbers. Under the usual measurability and Lipschitz continuity assumptions on G and F and also by the assumption that both G and F have square integrable representatives, so that

$$E\int_0^t R^2(s, x)dx := E_0^t[r_{s,x}(\Theta)]^2 ds + \mathcal{J}, \quad t \geq 0, \forall x \in \bar{\mathcal{C}} \tag{5.7}$$

is valid (R can be G or F, $R(s,x) = Cl[r_{s,x}(\Theta)]$), (5.6) has a unique continuous process solution X with values in $\bar{\mathcal{C}}$ (uniqueness is again in the sense of the weak equality in $\bar{\mathcal{C}}$). The solution process is a diffusion process with values in the generalized constant space.

Note that although the measurability and integrability concepts have not been made precise in the paper, the integral in (5.7) coincides with the representative integrability of [8].

6. Concluding Remarks

Although the research on the application of Colombeau's theory of non-linear distributions to stochastic analysis is far from accomplished, the fact that some new horizons have already started to emerge is quite apparent. In the first place the imbedding of the random linear distributions and Schwartz distribution-valued processes into a stochastic algebra (that of Colombeau) enables one to deal easily with singular objects, (e.g., $(\frac{dW}{dt})^2$ or any well-behaving non-linear function of the white noise).

On the other hand, an infinite dimensional version of Colombeau distributions constructed directly on abstract Wiener spaces offers great potential to provide efficient tools in stochastic analysis, comparable to those more sophisticated ones by Hida, Watanabe and Körezlioğlu and Üstünel ([12]). Also this relatively new theory has great versatility as far as the solutions to a nonlinear stochastic PDE is concerned, especially when the initial data is singular or distributional. When classical processes, random linear distributions or linear distribution-valued processes, fail to be solutions, some solution from the stochastic Colombeau algebras may be sought. Furthermore among the classical or linear distributional processes, the one nearest (in the sense of association relations discussed in the text) to the Colombeau solution can be worked out.

However the present theory is still at the stage of being rather more of an intellectual machinery than a tool that can be implemented in a large class of hard, concrete problems. There are needs for well established theories of general processes, stochastic integration and regular operations.

Note. As this volume was going to press, a recent survey by Oberguggenberger ([19]) came to the attention of the authors of the present paper. However the main emphases of the two papers go into quite different directions of the subject. [19] reviews also the approaches other than that of Colombeau in the deterministic theory of non-linear distributions. In the stochastic domain it concentrates only on the semi-linear wave equation discussed in [20] (c.f. (5.1), Section 5).

References

[1] A. Badrikian: Séminaire sur les Fonctions Aléatoire Linéaires et les Mesures Cylindrique, *Lecture Notes in Maths.* No. 379, Springer-Verlag,1974.

[2] H.A. Biagioni: A Non-linear Theory of Generalized Functions, *Lecture Notes in Maths.* No.1421, Springer-Verlag, 1990.

[3] J.F. Colombeau: New Generalized Functions and Multiplication of Distributions, *North Holland Math. Studies* **84** 1984.

[4] J.F. Colombeau: Elementary Introduction to New Generalized Functions, *North Holland Math. Studies* **113**, 1985.

[5] U. Çapar: An Introduction to Random Non-linear Distributions, *Abs. of the 22nd Conf. on Stochastic Processes and Their Applications*, pp.25, Vrije Universiteit, Amsterdam, Holland, 1993.

[6] U.Çapar: Random Generalized Fuctions Nonlinear in the Sense of Colombeau, *Proc. of the 6th Turkish National Math. Symposium*, Eastern Mediterranean University, North Cyprus, pp. 86–90, 1993.

[7] U. Çapar, H. Aktuğlu: On the Extension Problem in Processes Generalized in the Colombeau Sense, *Proc. of the 13th Prague Conference on Information Theory, Statistical Decision Functions and Random Processes*, Vol.1, pp. 70–77, Publishing House, Czech Acad. of Sci., 1998.

[8] U. Çapar, H. Aktuğlu: A New Construction of Random Colombeau Distributions, *Statistics and Probability Letters* **54** (2001), 291–299.

[9] U. Çapar, H. Aktuğlu: Topology, Duality and Stochasticity in Colombeau Related Algebras, *to appear.*

[10] Y. Egorov: A Contribution to the Theory of Generalized Functions, *Russian Math. Surveys* **45** (1990), 1–49.

[11] J.L. Kelley: General Topology, *Springer Verlag,* 1955.

[12] H. Körezlioğlu, A.S. Üstünel: A New Class of Distributions on Wiener Space, *Stoch. Analysis & Related Topics, Proc. of the 2nd Silivri Workshop* pp. 106–121, Springer Verlag, 1990.

[13] C. Martias: Stochastic Integration on General Function Spaces and its Applications, *Stochastic and Stochastics Reports* **57** (1996), 289–301.

[14] C. Martias: Colombeau's Generalized Functions on Abstract Wiener Space, *Université des Antilles et de la Guyane preprints.*

[15] C. Martias: Extension du Calcul d'Itô aux Distributions de Schwartz, *C.R. Acad. Sci. Paris* bf 321 (1995), 331–334.

[16] C. Martias: Random Non-linear Wave Equations, *Université des Antilles et de la Guyane preprints.*

[17] M. Oberguggenberger: Generalized Functions and Stochastic Processes, *Progress in Probability* **36** (1995), 215–229.

[18] M. Oberguggenberger: Non-linear Theories of Generalized Functions *(eds.)* S. Albeverio, W.A.J. Luxembourg, M.P.H. Wolff, *Advances in Analysis, Probability and Mathematical Physics, Contributions from Non-standard Analysis*, pp. 56–74, Kluwer, 1994.

[19] M. Oberguggenberger: Generalized Functions in Nonlinear Models — A Survey, *Nonlinear Analysis* **47** (2001), 5029–5040.

[20] M. Oberguggenberger, F. Russo: Non-linear SPDE's: Colombeau Solutions and Pathwise Limits, *Proc. of the 6th Oslo–Silivri Workshop on Stochastic Analysis and Related Topics*, 325–339, Birkhäuser, 1998.

[21] E.E. Rosinger: Distributions and Non-linear Partial Differential Equations, *Lecture Notes in Maths*, No. 684, Springer Verlag, 1978.

[22] E.E. Rosinger: Generalized Solutions of Non-linear Partial Differential Equations, *North Holland Math Studies* 146, 1987.

[23] A.S. Üstünel: Representation of Distributions on Wiener Space and Stochastic Calculus of Variation, *J.Funct. Anal. Vol.* **70** (1987), 126–139.

Department of Mathematics
Eastern Mediterranean University
G. Magusa
North Cyprus (via Mersin 10 – Turkey)

Progress in Probability, Vol. 53, 91–109

On Metrics for Tangent Processes on the Path Space

A.B. Cruzeiro and Kai-Nan Xiang

Abstract. Tangent processes, which are semimartingales with antisymmetric diffusion coefficient, were introduced in the framework of geometry on the path space by Cruzeiro and Malliavin ([3]). They correspond to an extension of the usual Cameron-Martin tangent space in Malliavin calculus, an extension which is in fact necessary in the non-flat situation. In this paper we discuss the possibility of introducing a metric or a Finsler structure on the space of tangent processes. We prove that the Levi–Civita and Cartan connections associated to the natural candidates to the Finsler geometry are not well defined.

1. Introduction

In this work M will denote an m-dimensional real C^∞ manifold, TM (T^*M) the tangent (cotangent) bundle of M, $O(M)$ the orthonormal frame bundle over M. Fix $r_0 \in O(M)$ with $\pi(r_0) = m_0$, π the canonical projection. Let A_i, $1 \le i \le m$, be the canonical horizontal vector fields on $O(M)$, and $x = (x^i(t))_{1\le i\le m}$ be the standard m-dimensional Brownian motion. Then the horizontal Brownian motion on $O(M)$ is defined by the following Stratonovich SDE:

$$d\tilde{p}(t) = \sum_{i=1}^{m} A_i(\tilde{p}(t)) \circ dx^i(t),\ \tilde{p}(0) = r_0. \tag{1.1}$$

Let

$$p = \pi \circ \tilde{p} = (\pi \circ \tilde{p}(t)). \tag{1.2}$$

We call the map $I : x \to p$ defined by (1.2) the Itô map.

We denote by $t^p_{\tau\leftarrow 0}$ the Itô parallel transport along the Brownian curve p on M from $p(0)$ to $p(\tau)$, and we assume M to be such that the Brownian curve is defined for all times.

Let

$$Q_\eta(\tau) = \int_0^\tau \Omega_{\tilde{p}(s)}(\eta(s), \circ dx(s)), \tag{1.3}$$

where η is an R^m-valued continuous adapted process with $\eta(0) = 0$,

$$\Omega_r(a, b)c = r^{-1} \circ R(ra, rb) \circ rc,\ \forall a, b, c \in R^m,\ r \in O(M),$$

$R(\cdot,\cdot)$ is the curvature tensor of M, and each $r \in O(M)$ is viewed as an isomorphism from R^m to $T_{\pi(r)}M$.

We consider the path spaces

$$P_{m_0}(M) := \{\gamma \in (C[0,1], M) | \gamma(0) = m_0 \in M\}$$

and $X = P_0(R^m)$. The Cameron–Martin space is the space H defined by

$$H = \{h \in P_0(R^m) \Big| \int_0^1 \left|\dot{h}(t)\right|^2 dt = \|h\|_H^2 < \infty\}.$$

Denote by I_* the differential of the Itô map I, then ([3], [4])

$$h(\tau) = h^*(\tau) - \int_0^\tau Q_{h^*(s)} \circ dx(s), \ \forall h \in H \tag{1.4}$$

where

$$h^*(\tau) = t^p_{0 \leftarrow \tau}(I_*(h)(\tau)).$$

As a consequence the Itô map is not Cameron–Martin differentiable. In other words, the tangent space of $P_{m_0}(M)$ at p given by

$$T_p(P_{m_0}(M)) = \{Z | Z(\tau) = t^p_{\tau \leftarrow 0}(z(\tau)), \ z \in H\}$$

is too small for I_* to provide an isomorphism from H to $T_p(P_{m_0}(M))$. Therefore Cruzeiro and Malliavin [3] (also see Driver [4]) introduced the space of tangent processes as follows.

A tangent process on X is an R^m-valued semimartingale such that

$$d\eta_\alpha(t) = a_{\alpha\beta}(t)dx^\beta(t) + \bar{c}_\alpha(t)dt = a_{\alpha\beta}(t) \circ dx^\beta(t) + c_\alpha(t)dt, \tag{1.5}$$

with

$$a_{\alpha\beta}(t) = -a_{\beta\alpha}(t), \ a_{\alpha\beta}(0) = 0.$$

The space of all such processes is denoted by $\mathcal{P}(X)$ and the tangent process space on $P_{m_0}(M)$ at p is given by

$$\mathcal{P}(P_{m_0}(M))(p) = \{\xi^* | \xi^*(\tau) = t^p_{\tau \leftarrow 0}(\xi(\tau)), \ \xi \in \mathcal{P}(X)\}.$$

Notice that the Itô map preserves the tangent process as an isomorphism and that the Lie bracket of tangent processes on $P_{m_0}(M)$ with some smoothness is closed ([3], [5]). To define a suitable metric on these tangent processes is a quite fundamental question that we discuss in the present paper. A suitable metric on tangent processes should, when restricted to the tangent space of $P_{m_0}(M)$, reduce to the Cameron–Martin space inner product. It should be convenient to determine the dual of each tangent process in order to compute easily the important geometric objects such as some intrinsic connections. Furthermore, it should embody some intrinsic geometric propositions of the probability space $(P_{m_0}(M), \mu)$ where μ is the Wiener measure on $P_{m_0}(M)$.

More concretely, in section 2, we calculate the Lie bracket of tangent processes; in section 3, we consider a Finsler geometry on $P_{m_0}(M)$; this Finsler geometry turns out to be Riemannian because the horizontal Brownian motion $\tilde{p}$ on $O(M)$ and the underlying Brownian motion p on M can be determined by each other (notice that, when considering smooth paths, the corresponding Finsler geometry frame is really Finslerian rather than Riemannian). Unfortunately the Levi–Civita connection or the Cartan connection coming from this Finsler geometry frame is not well defined (c.f. section 4). For a short review on finite-dimensional Finsler geometry, see the Appendix.

2. Lie Brackets of Tangent Processes

Let

$$d\xi_\alpha(t) = a_{\alpha\beta}(t) \circ dx^\beta(t) + c_\alpha(t)dt,\ d\xi'_\alpha(t) = a'_{\alpha\beta}(t) \circ dx^\beta(t) + c'_\alpha(t)dt$$

be two tangent processes on X. We shall use a superscript X to distinguish the derivatives on X from those on $P_{m_0}(M)$. For example, the directional derivative with respect to ξ on X is denoted by D^X_ξ. Throughout this paper, we assume the tangent processes to be smooth (for smoothness, we refer to [3] section 8).

Proposition 2.1 ([3]). *For a semimartingale η with $d\eta = a_\beta \circ dx^\beta + cdt$, and a process u, let*

$$\int u \emptyset d\eta = \int (ua_\beta) \circ dx^\beta + \int uc\ dt.$$

Then

$$\begin{aligned} a'_{\alpha\beta} \circ d\xi^\beta &= a'_{\alpha\beta} \emptyset d\xi^\beta; \\ d[D_\xi \xi']_\alpha &= [D_\xi(a'_{\alpha\beta})] \circ dx^\beta + (D_\xi c'_\alpha)dt + a'_{\alpha\beta} \circ d\xi^\beta \\ &= [D_\xi(a'_{\alpha\beta})] \circ dx^\beta + (D_\xi c'_\alpha)dt + a'_{\alpha\beta} \emptyset d\xi^\beta. \end{aligned}$$

Now we are in a position to calculate Lie brackets on $P_{m_0}(M)$. Let

$$\overline{\xi} = I_*^{-1}(\xi),\ \overline{\xi}' = I_*^{-1}(\xi').$$

Set

$$d\overline{\xi}_\alpha(t) = \overline{a}_{\alpha\beta}(t) \circ dx^\beta(t) + \overline{c}_\alpha(t)dt,\ d\overline{\xi}'_\alpha(t) = \overline{a}'_{\alpha\beta}(t) \circ dx^\beta(t) + \overline{c}'_\alpha(t)dt.$$

Therefore

$$d\overline{\xi} = d\xi - Q_\xi \circ dx,\ Q_\xi = \int_0^\cdot \Omega_{\tilde{p}(s)}(\xi(s), \circ dx(s)).$$

We see that

$$\overline{a}_{\alpha\beta} \circ dx^\beta + \overline{c}_\alpha d\tau = [a_{\alpha\beta} - \int_0^\tau (\Omega_{r\delta\alpha\beta}\xi^r) \circ dx^\delta(s)] \circ dx^\beta(\tau) + c_\alpha d\tau.$$

So we have shown that

$$\overline{a}_{\alpha\beta}(\tau) = a_{\alpha\beta}(\tau) - \int_0^\tau (\Omega_{r\delta\alpha\beta}\xi^r) \circ dx^\delta(s),\ \overline{c}_\alpha(\tau) = c_\alpha(\tau). \tag{2.1}$$

For a smooth cylinder function f on $P_{m_0}(M)$,

$$D_\xi D_{\xi'} f = \int_0^1 \sum_\alpha D_\xi(D_{\tau,\alpha} f) d\xi'_\alpha(\tau) + \int_0^1 \sum_\alpha D_{\tau,\alpha} f d(D_\xi \xi'_\alpha)(\tau),$$

where D_ξ, $D_{\tau,\alpha}$ are the directional derivatives on $P_{m_0}(M)$; combining with Main Lemma 6.3 in Cruzeiro and Malliavin [3], we obtain

Proposition 2.2. *We have*

$$d[\xi,\xi'] = d[[\xi,\xi']] + d(D_\xi\xi' - D_{\xi'}\xi)$$

where

$$d[[\xi,\xi']] = \Omega(\xi,\xi')(\circ dx) + [Q_\xi d\xi' - Q_{\xi'}d\xi].$$

Step 1. We compute $D_\xi\xi'$.

According to Proposition 2.1, we have

$$\begin{aligned} d(D_\xi\xi')_\alpha(\tau) &= d(D^X_{\bar{\xi}}\xi')_\alpha(\tau) \\ &= [D^X_{\bar{\xi}}(a'_{\alpha\beta})]\circ dx^\beta + [D^X_{\bar{\xi}}c'_\alpha]d\tau + a'_{\alpha\beta}\emptyset d\bar{\xi}^\beta \\ &= [D_\xi(a'_{\alpha\beta})]\circ dx^\beta + [D_\xi c'_\alpha]d\tau + a'_{\alpha\beta}\emptyset d\bar{\xi}^\beta \end{aligned}$$

and, from (2.1),

$$\begin{aligned} d(D_\xi\xi')(\tau) &= [D_\xi(a'_{\alpha\beta})]\circ dx^\beta + [D_\xi c'_\alpha]d\tau + a'_{\alpha\beta}c_\beta d\tau + \\ &\quad [a'_{\alpha\beta}a_{\beta r}(\tau) - a'_{\alpha\beta}\int_0^\tau(\Omega_{\lambda\delta\beta r}\xi^\lambda)\circ dx^\delta(s)]\circ dx^r(\tau). \end{aligned}$$

Step 2. We compute $D_\xi\xi' - D_{\xi'}\xi$.

$$\begin{aligned} d(D_\xi\xi' - D_{\xi'}\xi)_\alpha(\tau) &= [D_\xi(a'_{\alpha\beta}) - D_{\xi'}(a_{\alpha\beta})]\circ dx^\beta + [D_\xi c'_\alpha - D_{\xi'}c_\alpha]d\tau + \\ &\quad [a'_{\alpha\beta}a_{\beta r} - a_{\alpha\beta}a'_{\beta r}]\circ dx^r(\tau) + \\ &\quad [a_{\alpha\beta}\int_0^\tau \Omega_{\lambda\delta\beta r}\xi'^\lambda \circ dx^\delta(s)]\circ dx^r(\tau) - \\ &\quad [a'_{\alpha\beta}\int_0^\tau \Omega_{\lambda\delta\beta r}\xi^\lambda \circ dx^\delta(s)]\circ dx^r(\tau) + \\ &\quad [a'_{\alpha\beta}c_\beta - a_{\alpha\beta}c'_\beta]d\tau. \end{aligned}$$

Step 3. Consider $Q_\xi d\xi'$.

We have

$$(Q_\xi d\xi')_\alpha(\tau) = A_{\alpha\beta}(\tau)a'_{\beta r}(\tau)\circ dx^r(\tau) + A_{\alpha\beta}c'_\beta(\tau)d\tau,$$

where

$$\begin{aligned} A_{\alpha\beta}(\tau) &= \int_0^\tau \Omega_{\lambda\delta\alpha\beta}\xi^\lambda \circ dx^\delta, \\ (A'_{\alpha\beta}(\tau) &= \int_0^\tau \Omega_{\lambda\delta\alpha\beta}\xi'^\lambda \circ dx^\delta). \end{aligned}$$

Remark 2.3. We say that a tangent process ξ is of type A if $d\xi_\alpha(t) = a_{\alpha\beta}(t)\circ dx^\beta(t)$.

From steps 1 and 3, for

$$\begin{cases} d\xi^\alpha(\tau) = a_{\alpha\beta} \circ dx^\beta(\tau), \\ d\xi'^\alpha(\tau) = a'_{\alpha\beta} \circ dx^\beta(\tau), \end{cases}$$

$D_\xi\xi'(\cdot)+\int_0^\cdot Q_\xi d\xi'(t)$ is a tangent process of type A. Hence, the Markov Connection ∇ specified in Cruzeiro and Malliavin [3] can be extended naturally to tangent processes of type A as follows:

$$(\nabla_\xi\xi')(\tau) = D_\xi\xi'(\tau) + \int_0^\tau Q_\xi d\xi'(t).$$

Step 4. We compute $[Q_\xi d\xi' - Q_{\xi'}d\xi] + d(D_\xi\xi' - D_{\xi'}\xi)$.

$$\begin{aligned} &[Q_\xi d\xi' - Q_{\xi'}d\xi]_\alpha + d(D_\xi\xi' - D_{\xi'}\xi)_\alpha \\ &= [D_\xi(a'_{\alpha\beta}) - D_{\xi'}(a_{\alpha\beta})] \circ dx^\beta + \\ &\quad \{[D_\xi c'_\alpha - D_{\xi'}c_\alpha] + [a'_{\alpha\beta}c_\beta - a_{\alpha\beta}c'_\beta]\}d\tau \\ &\quad [a'_{\alpha\beta}a_{\beta r} - a_{\alpha\beta}a'_{\beta r}] \circ dx^r(\tau) + \\ &\quad \{[A_{\alpha\beta}a'_{\beta r} - a'_{\alpha\beta}A_{\beta r}] - [A'_{\alpha\beta}a_{\beta r} - a_{\alpha\beta}A'_{\beta r}]\} \circ dx^r(\tau) + \\ &\quad [A_{\alpha\beta}c'_\beta - A'_{\alpha\beta}c_\beta]d\tau. \end{aligned}$$

We hence obtain the following theorem.

Theorem 2.4.

$$\begin{aligned} d[\xi,\xi']_\alpha(\tau) &= \{\Omega_{\beta\delta\alpha r}\xi^\beta\xi'^\delta + \\ &\quad [A_{\alpha\beta}a'_{\beta r} - a'_{\alpha\beta}A_{\beta r}] - [A'_{\alpha\beta}a_{\beta r} - a_{\alpha\beta}A'_{\beta r}] + \\ &\quad [a'_{\alpha\beta}a_{\beta r} - a_{\alpha\beta}a'_{\beta r}] + [D_\xi(a'_{\alpha r}) - D_{\xi'}(a_{\alpha r})]\} \circ dx^r(\tau) + \\ &\quad \{[D_\xi c'_\alpha - D_{\xi'}c_\alpha] + [a'_{\alpha\beta}c_\beta - a_{\alpha\beta}c'_\beta] + \\ &\quad [A_{\alpha\beta}c'_\beta - A'_{\alpha\beta}c_\beta]\}d\tau. \end{aligned}$$

Corollary 2.5. *The Lie bracket on tangent processes of type A is closed.*

Indeed, if we consider

$$d\xi_\alpha(t) = a_{\alpha\beta}(t) \circ dx^\beta(t),\ d\xi'_\alpha(t) = a'_{\alpha\beta}(t) \circ dx^\beta(t),$$

then

$$\begin{aligned} d[\xi,\xi']_\alpha(\tau) &= \{\Omega_{\beta\delta\alpha r}\xi^\beta\xi'^\delta + \\ &\quad [A_{\alpha\beta}a'_{\beta r} - a'_{\alpha\beta}A_{\beta r}] - [A'_{\alpha\beta}a_{\beta r} - a_{\alpha\beta}A'_{\beta r}] + \\ &\quad [a'_{\alpha\beta}a_{\beta r} - a_{\alpha\beta}a'_{\beta r}] + [D_\xi(a'_{\alpha r}) - D_{\xi'}(a_{\alpha r})]\} \circ dx^r(\tau). \end{aligned}$$

3. An intrinsic Finsler geometry frame on $P_{m_0}(M)$

Denote the map $x \to \tilde{p}$ specified in (1.1) by $\tilde{I}$ and its differential by $\tilde{I}_*$.

Definition 3.1. The tangent space on $P_{r_0}(O(M))$ at $\tilde{p} \in P_{r_0}(O(M))$ is given by

$$T_{\tilde{p}}(P_{r_0}(O(M))) = \{H_h, \forall h = (h^i) \in H | H_h = h^i A_i\}.$$

The tangent process space on $P_{r_0}(O(M))$ is defined by

$$\mathcal{P}(P_{r_0}(O(M)))(\tilde{p}) = \tilde{I}_*(\mathcal{P}(X)).$$

Proposition 3.2 ([6]). *Let (θ, ω) be the canonical parallelism of $O(M)$ such that*

$$d\theta = -\omega \bigwedge \theta, \ d\omega = -\omega \bigwedge \omega + \Omega.$$

For any

$$\eta \in \mathcal{P}(X), \ d\eta_i(\tau) = a_{ij}(\tau) \circ dx^j(\tau) + c_i(\tau) d\tau,$$

let

$$\begin{cases} h_\eta := < \tilde{I}_*(\eta)(\tau), \theta > \in R^m, \\ v_\eta := < \tilde{I}_*(\eta)(\tau), \omega > \in so(m). \end{cases}$$

Then

$$\begin{cases} dh_\eta(\tau) = c(\tau) d\tau + (v_\eta(\tau) + a(\tau)) \circ dx(\tau), \\ dv_\eta(\tau) = \Omega(\circ dx(\tau), h_\eta(\tau)) \end{cases}$$

with $h_\eta(0) = 0, \ v_\eta(0) = 0$. Here $a = (a_{ij}), \ c = (c_i)$.

Definition 3.3. Decompose $\tilde{I}_*(\eta) \in \mathcal{P}(P_{r_0}(O(M)))$ into the following two parts:

$$\tilde{I}_*(\eta) = H_\eta + V_\eta$$

with

$$\begin{cases} < H_\eta(\tau), \theta >= h_\eta, \\ < H_\eta(\tau), \omega >= 0; \end{cases}$$

and

$$\begin{cases} < V_\eta(\tau), \theta >= 0, \\ < V_\eta(\tau), \omega >= v_\eta(\tau). \end{cases}$$

We call H_η and V_η the horizontal part and the vertical part of $\tilde{I}_*(\eta)$. Denote respectively the space of all such H_η and the family of all such V_η by $\mathcal{H}(P_{r_0}(O(M)))$ and $\mathcal{V}(P_{r_0}(O(M)))$; and call $\mathcal{H}(P_{r_0}(O(M)))$ the horizontal bundle and $\mathcal{V}(P_{r_0}(O(M)))$ the vertical bundle on $P_{r_0}(O(M))$ respectively.

Proposition 3.4. *Let π_* be the tangent map of $\pi : P_{r_0}(O(M)) \to P_{m_0}(M)$. Then*

$$\pi_*(\mathcal{V}(P_{r_0}(O(M)))) = 0, \ \pi_*(H_\eta) = I_*(\eta).$$

Therefore,

$$\mathcal{V}(P_{r_0}(O(M))) = ker\pi_*.$$

Definition 3.5 (The vertical lift map $l_{\tilde{p}}$). For

$$\tilde{p} \in P_{r_0}(O(M)), \ I_*(\eta) \in \mathcal{P}(P_{m_0}(M))(p), \ p = \pi \circ \tilde{p},$$

define

$$l_{\tilde{p}}(I_*\eta) = V_\eta \in \mathcal{V}(P_{r_0}(O(M))).$$

Then $l_{\tilde{p}}$ is a linear isomorphism from $\mathcal{P}(P_{m_0}(M))$ to $\mathcal{V}(P_{r_0}(O(M)))$ provided $\{r \in O(M) | \Omega_r = 0\}$ is zero-measurable with respect to the volume measure on $O(M)$.

Definition 3.6 (The horizontal map Θ). Define

$$\begin{aligned} \Theta \quad : \quad & \mathcal{V}(P_{r_0}(O(M))) \to \mathcal{H}(P_{r_0}(O(M))), \\ & V_\eta \qquad \to H_\eta. \end{aligned}$$

Then Θ is a linear isomorphism from $\mathcal{V}(P_{r_0}(O(M)))$ to $\mathcal{H}(P_{r_0}(O(M)))$ in the same situation as Definition 3.5.

Proposition 3.7.

$$\pi_* \circ \Theta_{\tilde{p}} = l_{\tilde{p}}^{-1}.$$

As in finite-dimensional Finsler geometry, Θ is called the horizontal map.

Proof. From $\pi_*(H_\eta) = I_*\eta$, we see that

$$\pi_* \circ \Theta_{\tilde{p}}(V_\eta) = I_*\eta = l_{\tilde{p}}^{-1}(V_\eta).$$

□

Definition 3.8 (The horizontal lift map $\mathcal{X}_{\tilde{p}}$). $\forall \tilde{p} \in P_{r_0}(O(M))$, define

$$\mathcal{X}_{\tilde{p}} = \Theta_{\tilde{p}} \circ l_{\tilde{p}}.$$

Clearly, $\mathcal{X}_{\tilde{p}}(I_*\eta) = H_\eta$.

Remark 3.9. So far we have defined the horizontal bundle on $P_{r_0}(O(M))$. Unfortunately we can not associate a nonlinear connection on $P_{m_0}(M)$ similarly to equation (A.2.4) in the Appendix.

In our case, the replacement of (A.2.4) would be

$$\tilde{D}\tilde{p} = l_{\tilde{p}}^{-1} \circ \kappa_{\tilde{p}} \circ \tilde{p}_*,$$

where $\tilde{p} = \tilde{I} \circ I^{-1}$, viewed as a map from $P_{m_0}(M)$ to $P_{r_0}(O(M))$; $\tilde{p}_*$ is the corresponding tangent map from $\mathcal{P}(P_{m_0}(M))$ to $\mathcal{P}(P_{r_0}(O(M)))$, and $\kappa_{\tilde{p}}$ is the projection map from $\mathcal{P}(P_{r_0}(O(M)))$ to $\mathcal{P}(P_{r_0}(O(M)))$.

For all $I_*\eta \in \mathcal{P}(P_{m_0}(M))$, by Proposition 3.2 and Definition 3.5, we have

$$\begin{aligned}\tilde{D}\tilde{p}(I_*\eta) &= l_{\tilde{p}}^{-1} \circ \kappa_{\tilde{p}} \circ \tilde{I}_* \circ I_*^{-1}(I_*\eta) \\ &= l_{\tilde{p}}^{-1} \circ \kappa_{\tilde{p}} \circ \tilde{I}_*(\eta) \\ &= l_{\tilde{p}}^{-1} V_\eta = I_*\eta.\end{aligned}$$

Definition 3.10 (Finsler-like structures). Assuming that $\{r \in O(M) | \Omega_r = 0\}$ is zero-measurable with respect to the volume measure on $O(M)$ and given an inner product $< \cdot, \cdot >$ on $\mathcal{V}(P_{r_0}(O(M)))$, define a nonnegative function F on $\mathcal{P}(P_{r_0}(O(M)))$ by

$$F(I_*\eta(p))^2 =< l_{\tilde{p}}(I_*\eta), l_{\tilde{p}}(I_*\eta) > .$$

Then, with the radial vertical vector field $l(u) = I_u(u)$ in the Appendix corresponding to $l_{\tilde{p}}(I_*\eta(p))$, we call F a Finsler-like structure or metric.

Remark 3.11. In finite-dimensional differential geometry, many geometric objects can be defined and calculated in local charts. While for infinite dimensional spaces, the definition using the local charts may be inconvenient; in such situations, the global approach can be taken. Here we take a global approach to the Cartan connection on the path space.

Definition 3.12 (The Cartan connection). Assume $\{r \in O(M) | \Omega_r = 0\}$ is zero-measurable with respect to the volume measure on $O(M)$. If F is a Finsler structure on $P_{m_0}(M)$, define a linear connection

$$\nabla : \mathcal{V}(P_{r_0}(O(M))) \times \mathcal{P}(P_{r_0}(O(M))) \to \mathcal{V}(P_{r_0}(O(M)))$$

such that for all $X, Y, Z \in \mathcal{V}(P_{r_0}(O(M)))$,

$$\begin{aligned}2 < \nabla_X Y, Z > &= X < Y, Z > + Y < X, Z > + < [Z, Y], X > - \\ &\quad < [X, Z], Y > - < [Y, X], Z > - Z < X, Y >;\end{aligned}$$

and for $X \in \mathcal{H}(P_{r_0}(O(M)))$ and $Y, Z \in \mathcal{V}(P_{r_0}(O(M)))$,

$$\begin{aligned}2 < \nabla_X Y, Z > = X < Y, Z > &+ (\Theta Y) < \Theta^{-1} X, Z > + < [Z, Y], \Theta^{-1} X > - \\ &< [\Theta^{-1} X, Z], Y > - < [Y, \Theta^{-1} X], Z > -(\Theta Z) < \Theta^{-1} X, Y > .\end{aligned}$$

From Remark A.4.4 in the Appendix, ∇ is called the Cartan connection.

Definition 3.13 (The nonlinear connection associated to the Cartan connection). Define

$$\tilde{\nabla} : \mathcal{P}(P_{m_0}(M)) \times \mathcal{P}(P_{m_0}(M)) \to \mathcal{P}(P_{m_0}(M))$$

by

$$l_{\tilde{p}}(\tilde{\nabla}_{I_*\eta} I_*\xi) = \nabla_{H_\eta} V_\xi, \ \forall \eta, \xi \in \mathcal{P}(X).$$

According to Proposition A.3.4, we call $\tilde{\nabla}$ the nonlinear connection.

Remark 3.14. Notice that when considering smooth paths, there is a corresponding Finsler geometry frame which is really Finslerian rather than Riemannian. But the above Finsler geometry frame on $P_{m_0}(M)$ is Riemannian because the horizontal Brownian motion $\tilde{p}$ on $O(M)$ and the underlying Brownian motion p on M can be determined by each other.

4. Metrics coming from the Finsler geometry frame

For three type A tangent processes ξ, ξ' and ξ'' with

$$\begin{cases} d\xi^\alpha(\tau) = a_{\alpha\beta} \circ dx^\beta(\tau), \\ d\xi'^\alpha(\tau) = a'_{\alpha\beta} \circ dx^\beta(\tau), \\ d\xi''^\alpha(\tau) = a''_{\alpha\beta} \circ dx^\beta(\tau); \end{cases}$$

we have

$$\begin{cases} d(v_\xi)_{r\delta}(\tau) = \Omega_{\beta\alpha r\delta}\xi^\alpha \circ dx^\beta(\tau), \\ d(v_{\xi'})_{r\delta}(\tau) = \Omega_{\beta\alpha r\delta}\xi'^\alpha \circ dx^\beta(\tau), \\ d(v_{\xi''})_{r\delta}(\tau) = \Omega_{\beta\alpha r\delta}\xi''^\alpha \circ dx^\beta(\tau). \end{cases}$$

According to Remark 3.14 and Definition 3.10, we define an inner product on type A tangent processes as follows.

$$< \xi, \xi' > = < v_\xi, v_{\xi'} > = \sum_{\beta,r,\delta} \int_0^1 \Omega^2_{\alpha\beta r\delta}\xi^\alpha \xi'^\alpha d\tau.$$

We now compute the Levi-Civita connection ∇ associated to this L^2-metric. Notice that

$$\begin{aligned} 2 < \nabla_{\xi''}\xi, \xi' > &= D_{\xi''} < \xi, \xi' > + D_\xi < \xi'', \xi' > + < [\xi', \xi], \xi'' > - \\ &\quad < [\xi'', \xi'], \xi > - < [\xi, \xi''], \xi' > - D_{\xi'} < \xi'', \xi > . \end{aligned}$$

Step 1. We compute $D_{\xi''} < \xi, \xi' >$.

$$\begin{aligned} D_{\xi''} < \xi, \xi' > &= \sum_{\beta,r,\delta} D_{\xi''}[\int_0^1 \Omega^2_{\alpha\beta r\delta}\xi^\alpha\xi'^\alpha d\tau] = \sum_{\beta,r,\delta}\int_0^1 D_{\xi''}[\Omega^2_{\alpha\beta r\delta}\xi^\alpha\xi'^\alpha]d\tau \\ &= 2\sum_{\beta,r,\delta}\int_0^1 \Omega_{\alpha\beta r\delta}(D_{\xi''}\Omega_{\alpha\beta r\delta})\xi^\alpha\xi'^\alpha d\tau + \\ &\quad \sum_{\beta,r,\delta}\int_0^1 \Omega^2_{\alpha\beta r\delta}[D_{\xi''}\xi^\alpha]\xi'^\alpha d\tau + \\ &\quad \sum_{\beta,r,\delta}\int_0^1 \Omega^2_{\alpha\beta r\delta}\xi^\alpha[D_{\xi''}\xi'^\alpha]d\tau. \end{aligned}$$

$$
\begin{aligned}
2 < \nabla_{\xi''}\xi, \xi' > = {} & 2\sum_{\beta,r,\delta}\int_0^1 \Omega_{\alpha\beta r\delta}(D_{\xi''}\Omega_{\alpha\beta r\delta})\xi^{\alpha}\xi'^{\alpha}d\tau + \\
& 2\sum_{\beta,r,\delta}\int_0^1 \Omega_{\alpha\beta r\delta}(D_{\xi}\Omega_{\alpha\beta r\delta})\xi''^{\alpha}\xi'^{\alpha}d\tau - \\
& 2\sum_{\beta,r,\delta}\int_0^1 \Omega_{\alpha\beta r\delta}(D_{\xi'}\Omega_{\alpha\beta r\delta})\xi''^{\alpha}\xi^{\alpha}d\tau + \\
& \sum_{\beta',r',\delta'}\int_0^1 \Omega^2_{\alpha\beta' r'\delta'}\int_0^{\tau}\{\Omega_{\beta\delta\alpha r}\xi'^{\beta}\xi^{\delta} + A'_{\alpha\beta}a_{\beta r} - \\
& A_{\alpha\beta}a'_{\beta r}\}\circ dx^r(t)\xi''^{\alpha}(\tau)d\tau - \\
& \sum_{\beta',r',\delta'}\int_0^1 \Omega^2_{\alpha\beta' r'\delta'}\int_0^{\tau}\{\Omega_{\beta\delta\alpha r}\xi''^{\beta}\xi'^{\delta} + A''_{\alpha\beta}a'_{\beta r} - \\
& A'_{\alpha\beta}a''_{\beta r}\}\circ dx^r(t)\xi^{\alpha}(\tau)d\tau - \\
& \sum_{\beta',r',\delta'}\int_0^1 \Omega^2_{\alpha\beta' r'\delta'}\int_0^{\tau}\{\Omega_{\beta\delta\alpha r}\xi^{\beta}\xi''^{\delta} + A_{\alpha\beta}a''_{\beta r} - \\
& A''_{\alpha\beta}a_{\beta r}\}\circ dx^r(t)\xi'^{\alpha}(\tau)d\tau \\
= {} & 2\sum_{\beta,r,\delta}\int_0^1 \Omega_{\alpha\beta r\delta}(D_{\xi''}\Omega_{\alpha\beta r\delta})\xi^{\alpha}\xi'^{\alpha}d\tau + \\
& 2\sum_{\beta,r,\delta}\int_0^1 \Omega_{\alpha\beta r\delta}(D_{\xi}\Omega_{\alpha\beta r\delta})\xi''^{\alpha}\xi'^{\alpha}d\tau - \\
& 2\sum_{\beta,r,\delta}\int_0^1 \Omega_{\alpha\beta r\delta}(D_{\xi'}\Omega_{\alpha\beta r\delta})\xi''^{\alpha}\xi^{\alpha}d\tau + \\
& \sum_{\beta',r',\delta'}\int_0^1 \Omega^2_{\alpha\beta' r'\delta'}\int_0^{\tau}\{\Omega_{\beta\delta\alpha r}\xi'^{\beta}\xi^{\delta} + A'_{\alpha\beta}a_{\beta r}\}\circ dx^r(t)\xi''^{\alpha}(\tau)d\tau - \\
& \sum_{\beta',r',\delta'}\int_0^1 \Omega^2_{\alpha\beta' r'\delta'}\int_0^{\tau} A_{\alpha\beta}\circ d\xi'^{\beta}(t)\xi''^{\alpha}(\tau)d\tau - \\
& \sum_{\beta',r',\delta'}\int_0^1 \Omega^2_{\alpha\beta' r'\delta'}\int_0^{\tau}\{\Omega_{\beta\delta\alpha r}\xi''^{\beta}\xi'^{\delta} - A'_{\alpha\beta}a''_{\beta r}\}\circ dx^r(t)\xi^{\alpha}(\tau)d\tau - \\
& \sum_{\beta',r',\delta'}\int_0^1 \Omega^2_{\alpha\beta' r'\delta'}\int_0^{\tau} A''_{\alpha\beta}\circ d\xi'^{\beta}(t)\xi^{\alpha}(\tau)d\tau - \\
& \sum_{\beta',r',\delta'}\int_0^1 \Omega^2_{\alpha\beta' r'\delta'}\int_0^{\tau}\{\Omega_{\beta\delta\alpha r}\xi^{\beta}\xi''^{\delta} + A_{\alpha\beta}a''_{\beta r} - \\
& A''_{\alpha\beta}a_{\beta r}\}\circ dx^r(t)\xi'^{\alpha}(\tau)d\tau.
\end{aligned}
$$

Therefore we have shown the following:

Theorem 4.1.

$$
\begin{aligned}
2 < \nabla_{\xi''}\xi, \xi' > = {} & 2\sum_{\beta,r,\delta}\int_0^1 \Omega_{\alpha\beta r\delta}(D_{\xi''}\Omega_{\alpha\beta r\delta})\xi^\alpha\xi'^\alpha d\tau + \\
& 2\sum_{\beta,r,\delta}\int_0^1 \Omega_{\alpha\beta r\delta}(D_{\xi}\Omega_{\alpha\beta r\delta})\xi''^\alpha\xi'^\alpha d\tau - \\
& 2\sum_{\beta,r,\delta}\int_0^1 \Omega_{\alpha\beta r\delta}(D_{\xi'}\Omega_{\alpha\beta r\delta})\xi''^\alpha\xi^\alpha d\tau + \\
& \sum_{\beta',r',\delta'}\int_0^1 \Omega^2_{\alpha\beta' r'\delta'}\int_0^\tau \{\Omega_{\beta\delta\alpha r}\xi'^\beta\xi^\delta\} \circ dx^r(t)\xi''^\alpha(\tau)d\tau + \\
& \sum_{\beta',r',\delta'}\int_0^1 \Omega^2_{\alpha\beta' r'\delta'}\int_0^\tau A'_{\alpha\beta} \circ d\xi^\beta(t)\xi''^\alpha(\tau)d\tau - \\
& \sum_{\beta',r',\delta'}\int_0^1 \Omega^2_{\alpha\beta' r'\delta'}\int_0^\tau A_{\alpha\beta} \circ d\xi'^\beta(t)\xi''^\alpha(\tau)d\tau - \\
& \sum_{\beta',r',\delta'}\int_0^1 \Omega^2_{\alpha\beta' r'\delta'}\int_0^\tau \{\Omega_{\beta\delta\alpha r}\xi''^\beta\xi'^\delta\} \circ dx^r(t)\xi^\alpha(\tau)d\tau + \\
& \sum_{\beta',r',\delta'}\int_0^1 \Omega^2_{\alpha\beta' r'\delta'}\int_0^\tau A'_{\alpha\beta} \circ d\xi''^\beta(t)\xi^\alpha(\tau)d\tau - \\
& \sum_{\beta',r',\delta'}\int_0^1 \Omega^2_{\alpha\beta' r'\delta'}\int_0^\tau A''_{\alpha\beta} \circ d\xi'^\beta(t)\xi^\alpha(\tau)d\tau - \\
& \sum_{\beta',r',\delta'}\int_0^1 \Omega^2_{\alpha\beta' r'\delta'}\int_0^\tau \{\Omega_{\beta\delta\alpha r}\xi^\beta\xi''^\delta\} \circ dx^r(t)\xi'^\alpha(\tau)d\tau - \\
& \sum_{\beta',r',\delta'}\int_0^1 \Omega^2_{\alpha\beta' r'\delta'}\int_0^\tau A_{\alpha\beta} \circ d\xi''^\beta(t)\xi'^\alpha(\tau)d\tau + \\
& \sum_{\beta',r',\delta'}\int_0^1 \Omega^2_{\alpha\beta' r'\delta'}\int_0^\tau A''_{\alpha\beta} \circ d\xi^\beta(t)\xi'^\alpha(\tau)d\tau.
\end{aligned}
$$

Similarly, the following theorem holds.

Theorem 4.2. *Defining*

$$
< \xi, \xi' >= \sum_{\beta,r,\delta}\int_0^1 \xi^\alpha\xi'^\alpha d\tau,
$$

we have:

$$\begin{aligned}2 < \nabla_{\xi''}\xi, \xi' > \; = \; & \int_0^1 \int_0^\tau \{\Omega_{\beta\delta\alpha r}\xi'^{\beta}\xi^{\delta}\} \circ dx^r(t)\xi''^{\alpha}(\tau)d\tau + \\ & \int_0^1 \int_0^\tau A'_{\alpha\beta} \circ d\xi^{\beta}(t)\xi''^{\alpha}(\tau)d\tau - \\ & \int_0^1 \int_0^\tau A_{\alpha\beta} \circ d\xi'^{\beta}(t)\xi''^{\alpha}(\tau)d\tau - \\ & \int_0^1 \int_0^\tau \{\Omega_{\beta\delta\alpha r}\xi''^{\beta}\xi'^{\delta}\} \circ dx^r(t)\xi^{\alpha}(\tau)d\tau + \\ & \int_0^1 \int_0^\tau A'_{\alpha\beta} \circ d\xi''^{\beta}(t)\xi^{\alpha}(\tau)d\tau - \\ & \int_0^1 \int_0^\tau A''_{\alpha\beta} \circ d\xi'^{\beta}(t)\xi^{\alpha}(\tau)d\tau - \\ & \int_0^1 \int_0^\tau \{\Omega_{\beta\delta\alpha r}\xi^{\beta}\xi''^{\delta}\} \circ dx^r(t)\xi'^{\alpha}(\tau)d\tau - \\ & \int_0^1 \int_0^\tau A_{\alpha\beta} \circ d\xi''^{\beta}(t)\xi'^{\alpha}(\tau)d\tau + \\ & \int_0^1 \int_0^\tau A''_{\alpha\beta} \circ d\xi^{\beta}(t)\xi'^{\alpha}(\tau)d\tau.\end{aligned}$$

Corollary 4.3. *In the situations of Theorems* 4.1–4.2, *the Levi-Civita connections are not well defined.*

Proof. It suffices to prove the corollary in the case of Theorem 4.2. Assume the converse, namely the existence of $\eta \in \mathcal{P}(X)$ of type A such that

$$2 < \nabla_{\xi''}\xi, \xi' > = < \eta, \xi' > \tag{4.1}$$

for any type A process $\xi' \in \mathcal{P}(X)$. Then by the Hahn–Banach extension theorem, (4.1) determines a continuous linear functional on $L^2[0,1]$. Namely, (4.1) holds for any $\xi' \in L^2[0,1]$.

For any $\sigma \in (0,1)$, let

$$e_{\sigma,\alpha}(\tau) = I_{[\sigma,1]}(\tau)e_\alpha, \ \forall \tau \in [0,1],$$

where e_α is the α-th unit vector of R^m. Replacing ξ' by $e_{\sigma,\alpha}$ in (4.1), we can obtain that $\eta \notin \mathcal{P}(X)$. □

Remark 4.4. For the $H^{\frac{1}{2}-\epsilon}$ $(0 < \epsilon < \frac{1}{2})$ metric and the following two metrics, the same conclusion of Corollary 4.3 holds (by a similar reason).

$$< \xi, \xi' > = \sum_{r,\delta} \int_0^1 \int_0^\tau \Omega_{\alpha\beta r\delta}\xi^{\alpha} \circ dx^{\beta}(t)\Omega_{\alpha\beta r\delta}\xi'^{\alpha} \circ dx^{\beta}(t)d\tau$$

or

$$< \xi, \xi' >= \sum_{\alpha,\beta} \int_0^1 a_{\alpha\beta} a'_{\alpha\beta} d\tau.$$

Appendix A. An Introduction to Finsler Geometry

Let $(x^1, \cdots, x^m) = (x^i) : U \to R^m$ be a local coordinate system on an open subset $U \subset M$. For any $x \in U$, $u \in T_x M$, we have $u = u^a \frac{\partial}{\partial x^a}$. Here and henceforth we use the Einstein convention. Then a local coordinate system on TM and a local frame of $T(TM)$ can be given respectively by

$$(x, u) := (x^1, \cdots, x^m; u^1, \cdots, u^m) \text{ and } \{\partial_1, \cdots, \partial_m; \dot{\partial}_1, \cdots, \dot{\partial}_m\},$$

where $\partial_a = \frac{\partial}{\partial x^a}$ and $\dot{\partial}_b = \frac{\partial}{\partial u^b}$.

A.1. Finsler Structures (Metrics)

We denote by $\tilde{M} := TM \backslash \{0\}$, 0 being the zero section.

Definition A.1.1. A Finsler structure (metric) of M is a function $F : TM \to [0, \infty)$ such that

(i) F is C^∞ on $\tilde{M}$;

(ii) $F(x, \lambda u) = \lambda F(x, u)$, $\forall \lambda \in R^1$, $x \in M$, $u \in T_x M$;

(iii) The Hessian matrix $(g_{ij}(x, u)) := ([\frac{1}{2}F^2]_{u^i u^j})$ is positive definite at each point of $\tilde{M}$.

Remark A.1.2. (i) F is smooth on TM if and only if F is a Riemannian metric.

(ii) In Definition A.1 (iii), the partial derivatives of $\frac{1}{2}F^2$ are taken with respect to the u^i's; one can check that the positive definiteness of the Hessian matrix is independent of the choice of the local coordinate system of TM.

Proposition A.1.3.

$$F(x, u)^2 = g_{ij}(x, u) u^i u^j,$$
$$g_{ij}(x, \lambda u) = g_{ij}(x, u), \forall \lambda \in R^1 \backslash \{0\}.$$

Remark A.1.4. From Proposition A.3, $g_{ij}(x, u)$ is defined for all $(x, u) \in \tilde{M}$, and is invariant under nonzero rescaling in u. When $(g_{ij}(x, u))$ does not depend on $u \neq 0$, then $F(x, u)$ behaves exactly like an inner product on $T_x M$. However, Proposition A.3 shows that on the line $((x, \lambda u)_{\lambda \in R^1})$, $F(x, u)$ is just an inner product. In other words, F is Riemannian on each line passing through 0 in $T_x M$, $\forall x \in M$.

The Finsler structure F can be lifted to be a Riemannian structure. This will be done in Section A.4.

A.2. Preliminaries I: nonlinear connections

Definition A.2.1. The vertical bundle over TM is the bundle

$$\tilde{\pi} : \mathcal{V} \to TM,$$

where $\mathcal{V} = ker(\pi_*) \subset T(TM)$.

Let $j_x : T_xM \to TM$ be the inclusion map, and $i_u : T_xM \to T_u(T_xM)$ be the usual identification. A natural isomorphism is obtained as follows.

$$l_u = (j_{\pi(u)})_* \circ i_u : T_{\pi(u)}M \to \mathcal{V}_u. \tag{A.2.1}$$

Definition A.2.2. The radial vertical vector field $l : TM \to \mathcal{V}$ is determined by

$$l(u) = l_u(u) \in \mathcal{V}_u.$$

Locally, for $u = u^a \frac{\partial}{\partial x^a} \in T_{\pi(u)}M$, we have

$$l_u(\frac{\partial}{\partial x^j}|_{\pi(u)}) = \dot{\partial}_j|_u; \; l(u) = l(u^a \frac{\partial}{\partial x^a}) = u^a \dot{\partial}_a|_u.$$

Definition A.2.3. A horizontal bundle $\mathcal{H}$ over TM is a bundle such that

$$T(TM) = \mathcal{H} \bigoplus \mathcal{V}.$$

Definition A.2.4. A horizontal map $\Theta : \mathcal{V} \to T(TM)$ is a map satisfying

$$\forall u \in TM, \; (\pi_* \circ \Theta)_u = l_u^{-1}.$$

For any $v = v^a \frac{\partial}{\partial x^a} \in T_xM, \; \xi = \xi^b \frac{\partial}{\partial x^b}$, one has

$$d\xi_x(v) = v^a \partial_a|_u + v^a \frac{\partial \xi^b}{\partial x^a}(x) \dot{\partial}_b|_u. \tag{A.2.2}$$

Denote by $\mathcal{X}(E)$ the space of sections of E, where E is a bundle over M.

For $\xi, \xi' \in \mathcal{X}(TM), \; \xi(x) = \xi'(x) = u$, we have

$$d\xi'_x - d\xi_x : T_xM \to \mathcal{V}_u.$$

Definition A.2.5. We call $\tilde{D} : \mathcal{X}(TM) \to \mathcal{X}(T^*M \bigotimes TM)$ a nonlinear connection if

$$\tilde{D}0 = 0, \; \tilde{D}\xi_x - \tilde{D}\xi'_x = l_u^{-1} \circ (d\xi_x - d\xi'_x), \tag{A.2.3}$$

where $u \in T_xM, \; \xi(x) = \xi'(x) = u$.

Let $\mathcal{H}$ be a horizontal bundle, $T\tilde{M} = \mathcal{H} \bigoplus \mathcal{V}$; and $\kappa : T\tilde{M} \to \mathcal{V}$ the vertical projection. Define a nonlinear connection on M by

$$\tilde{D}_{\mathcal{H}}\xi_x = l_{\xi(x)}^{-1} \circ \kappa_{\xi(x)} \circ d\xi_x, \; \forall x \in M, \; \xi \in \mathcal{X}(TM). \tag{A.2.4}$$

For a nonlinear connection $\tilde{D}$, $\forall u \in \tilde{M}, \; \xi \in \mathcal{X}(TM), \; \xi(x) = u$, define a horizontal map $\Theta_u^{\tilde{D}} : \mathcal{V}_u \to T_u\tilde{M}$ as follows:

$$\Theta_u^{\tilde{D}} = d\xi_x \circ l_u^{-1} - l_u \circ \tilde{D}\xi_x \circ l_u^{-1}. \tag{A.2.5}$$

Given a horizontal map Θ, a horizontal bundle $\mathcal{H}^\Theta$ is determined by

$$\mathcal{H}_u^\Theta = \Theta_u(\mathcal{V}_u). \tag{A.2.6}$$

Definition A.2.6.
$$\mathcal{H} \to \tilde{D}_{\mathcal{H}},\ \tilde{D} \to \Theta^{\tilde{D}},\ \Theta \to \mathcal{H}^{\Theta}$$
define one-to-one correspondences.

Definition A.2.7. The radial horizontal vector field $\mathcal{X} : TM \to \mathcal{H}^{\Theta}$ is determined by
$$\mathcal{X} = \Theta \circ l \in \mathcal{X}(\mathcal{H}^{\Theta}).$$
For $u \in T_x M$, set
$$\mathcal{X}_u = \Theta \circ l_u : T_x M \to \mathcal{H}^{\Theta}_u.$$

A.3. Preliminaries II: Vertical Connections

Definition A.3.1. A vertical connection $D : \mathcal{X}(\mathcal{V}) \to \mathcal{X}(T^*\tilde{M} \bigotimes \mathcal{V})$ satisfies
$$D(fV) = fDV + df \bigotimes V,$$
for any $f \in C^{\infty}(\tilde{M}),\ V \in \mathcal{X}(\mathcal{V})$. Let $\nabla_X V = (DV)(X)$.

Let D be a vertical connection, and denote by $\bigwedge : T\tilde{M} \to \mathcal{V}$ the bundle map defined by
$$\bigwedge(X) = \nabla_X l,$$
here l is given in Definition A.6.

Definition A.3.2. D is "good" if $\bigwedge|_{\mathcal{V}} : \mathcal{V} \to \mathcal{V}$ is a bundle isomorphism.

Proposition A.3.3.
$$D \text{ is "good"} \iff T\tilde{M} = ker \bigwedge \bigoplus \mathcal{V} \iff \mathcal{H} = ker \bigwedge \text{ is a horizontal bundle.}$$

Let D be a "good" vertical connection, $\Theta : \mathcal{V} \to \mathcal{H}$ denote the horizontal map associated to the horizontal bundle $\mathcal{H} = ker \bigwedge$. Using Θ, we may define a linear connection (still denoted by D) on $\mathcal{H}$:
$$\forall H \in \mathcal{X}(\mathcal{H}),\ DH = \Theta(D\Theta^{-1}(H)),$$
namely,
$$\forall H \in \mathcal{X}(\mathcal{H}),\ X \in T\tilde{M},\ \nabla_X H = \Theta(\nabla_X(\Theta^{-1}H)).$$
Now a linear connection D on $\tilde{M}$ can be defined as follows.
$$\forall Y \in \mathcal{X}(T\tilde{M}),\ DY = D(\kappa Y) + D(Y - \kappa Y),$$
that is
$$\forall X, Y \in \mathcal{X}(T\tilde{M}),\ \nabla_X Y = \nabla_X(\kappa Y) + \nabla_X(Y - \kappa Y),$$
where $\kappa : T\tilde{M} \to \mathcal{V}$ is the vertical projection.

By definition,
$$\forall H \in \mathcal{H},\ V \in \mathcal{V},\ \nabla_H \mathcal{X} = 0,\ \nabla_V \mathcal{X} = \Theta \bigwedge(V).$$

For all $\xi \in \mathcal{X}(TM)$, let
$$\xi^{\mathcal{V}}(u) = l_u(\xi(\pi(u)) \in \mathcal{V}_u,\ \xi^{\mathcal{H}}(u) = \mathcal{X}_u(\xi(\pi(u)) \in \mathcal{H}_u.$$

Proposition A.3.4. *Let $\tilde{D}$ be the nonlinear connection associated to $\mathcal{H}$. Take $\xi \in \mathcal{X}(TM)$, $x \in M$. Then*

$$\forall u \in T_x M, \; \mathcal{X}_{\xi(x)}(\tilde{\nabla}_u \xi) = \nabla_{\mathcal{X}_{\xi(x)}(u)} \xi^{\mathcal{H}},$$

that is

$$\mathcal{X}_{\xi(x)} \circ \tilde{D}\xi = D\xi^{\mathcal{H}} \circ \mathcal{X}_{\xi(x)}.$$

Definition A.3.5 (Torsion). Let D be the linear connection on $\tilde{M}$ induced by a "good" vertical connection. The torsion tensor θ associated to D is defined by

$$\theta(X,Y) = \nabla_X Y - \nabla_Y X - [X,Y], \; \forall X, Y \in \mathcal{X}(T\tilde{M}).$$

A.4. A Riemannian metric on the vertical bundle of $\tilde{M}$

In this section, we associate to a Finsler structure F a Riemannian metric on $\mathcal{V}$ and a "good" vertical connection which is the generalization of the Levi-Civita connection.

From Definition A.1 (iii), we can introduce a Riemannian structure on $\mathcal{V}$:

$$\forall V, W \in \mathcal{V}_u, \; x = \pi(u), \; < V, W >_u = g_{ab}(x,u) V^a W^b. \qquad (A.4.1)$$

Notice that

$$F(x,u)^2 = < l(u), l(u) > . \qquad (A.4.2)$$

Theorem A.4.1. *There is a unique vertical connection (called the Cartan connection) $D : \mathcal{X}(\mathcal{V}) \to \mathcal{X}(T^*\tilde{M} \bigotimes \mathcal{V})$ such that*

(i) D is "good";

(ii) $\forall X \in T\tilde{M}, \; V, W \in \mathcal{X}(\mathcal{V})$,

$$X < V, W > = < \nabla_X V, W > + < V, \nabla_X W >;$$

(iii) the torsion θ of the linear connection on $\tilde{M}$ induced by D satisfies

$$\theta(V, W) = 0, \; \forall V, W \in \mathcal{V};$$

and

(iv)

$$\theta(X,Y) \in \mathcal{V}, \; \forall X, Y \in \mathcal{H}.$$

Remark A.4.2. For the Cartan connection D,

$$\theta(X,Y) \in \mathcal{V}, \; \forall X, Y \in T\tilde{M} \Longleftrightarrow F \text{ is Riemannian.}$$

In this case, the nonlinear connection associated to D coincides with the Levi-Civita connection.

Remark A.4.3. We can define a Riemannian metric on the whole $T\tilde{M}$. For all $X, Y \in \mathcal{H}$, let

$$< X, Y > = < \Theta^{-1}(X), \Theta^{-1}(K) > .$$

Then define a Riemannian metric on $T\tilde{M}$ just by stating that $\mathcal{H}$ is orthogonal to $\mathcal{V}$.

It is easy to check that

$$\forall X, Y, Z \in \mathcal{X}(T\tilde{M}),\ \ X < Y, Z >=< \nabla_X Y, Z > + < Y, \nabla_X Z > .$$

Remark A.4.4. One can show that for all $X, Y, Z \in \mathcal{V}$,

$$\begin{aligned} 2 < \nabla_X Y, Z > \ &= \ X < Y, Z > + Y < X, Z > + < [Z, Y], X > - \\ & < [X, Z], Y > - < [Y, X], Z > - Z < X, Y >; \end{aligned}$$

and for $X \in \mathcal{H}$ and $Y, Z \in \mathcal{V}$,

$$\begin{aligned} 2 < \nabla_X Y, Z > = X < Y, Z > &+ (\Theta Y) < \Theta^{-1} X, Z > + < [Z, Y], \Theta^{-1} X > - \\ & < [\Theta^{-1} X, Z], Y > - < [Y, \Theta^{-1} X], Z > - (\Theta Z) < \Theta^{-1} X, Y > . \end{aligned}$$

References

[1] M. Abate and G. Patrizio, "Finsler metric-A global approach," Lect. Note. Math. **1591**, Springer-Verlag.

[2] D. Bao, S.S. Chern and Z. Shen, "An introduction to Riemann–Finsler geometry," Springer-Verlag, 1999.

[3] A.B. Cruzeiro and P. Malliavin, Renormalized differential geometry on path spaces: structural equation, curvature, J. Funct. Anal. **139** (1996), 119–181.

[4] B. Driver, A Cameron–Martin type quasi-invariance theorem for Brownian motion on a compact manifold, J.Funct.Anal. **110** (1992), 272–376..

[5] B. Driver, The Lie bracket of adapted vector fields on Wiener spaces, Appl. Math. Optimal. **39** **(2)** (1999), 179–210.

[6] S. Fang and P. Malliavin, Stochastic analysis on the path space of a Riemannian manifold, J. Funct. Anal. **118** (1993), 249–274.

Dep. Matemática I.S.T. and Grupo de Física-Matemática U.L.
Av. Rovisco Pais
1059-001 Lisboa
Portugal
e-mail: `cruzeiro@math.ist.utl.pt`

Department of Probability and Statistics
Peking University
Beijing 100871
P.R. of China
and
Grupo de Física-Matemática U.L.
1649-003 Lisboa
Portugal

Progress in Probability, Vol. 53, 111–128

Weak Approximation for Semilinear Stochastic Evolution Equations

Erika Hausenblas

Abstract. We investigate the accuracy of approximation of $\mathbb{E}[\phi(X_t)]$, where X_t is a solution of a stochastic partial differential equation, and $\varphi : \tilde{E} \to \mathbb{R} \in L^\infty(\tilde{E})$, $\tilde{E}$ is a finite dimensional subspace of E. For the approximation we consider the implicit Euler scheme, and the Cranc–Nicholson scheme. Since the explicit Euler scheme is conditionally stable, we consider in this case discretization in space and time. As in the finite dimensional case it turns out that the rate of convergence can be improved to order 1.

1. Introduction

The stochastic calculus of variation was developed by Paul Malliavin in 1976 in order to provide a probabilistic proof of Hörmander's theorem, which has since come to be known as the Malliavin calculus. One of the features is the integration by parts formula, which allows us to study solutions of stochastic differential equations as a function of the forward variable. This together with some estimates of the backward variable allows us to deal with the weak error for numerical approximation of stochastic partial differential equations, which is our point of interest.

Let E be a certain Hilbert space. We consider the nonlinear stochastic partial differential equation

$$\begin{cases} dX_t &= (AX_t + f(X_t))\, dt + Q^{\frac{1}{2}} d\omega(t), \\ X_0 &= x \in E, \end{cases}$$

where A is a certain differential operator satisfying an ellipticity condition, and $\omega(t)$ an E-valued cylindrical Wiener process carried by the stochastic basis $(\Omega, \mathcal{F}, P)$ with filtration $(\mathcal{F}_t)_{t\geq 0}$, Q an operator on E.

Our point of interest is the accuracy of computing the entity $\mathbb{E}[\varphi(X_T)]$ at a time T by a Monte-Carlo algorithm. Here we consider $\varphi : \tilde{E} \to \mathbb{R} \in L^\infty(\tilde{E})$, where $\tilde{E}$ is a finite dimensional subspace. This entity is often of interest in, e.g., linear filtering, controlled reaction–diffusion equations, dealing with potential theory (see Mavrantzas and Öttinger [11]), or evaluating e.g., $\mathbb{P}(|X_T|_{\tilde{E}} > \alpha)$ for a given $\alpha \in \mathbb{R}^+$.

In Monte-Carlo simulation, a large number M of independent trajectories of X_T^i are simulated on a computer. The entity $\mathbb{E}[\varphi(X_T)]$ is approximated by

$$\frac{1}{M}\sum_i^M \varphi(X_T^i).$$

The resulting error depends on the choice of the approximate process and the two parameters M and n. The effects of M can be described by the Central Limit Theorem. The effects of n can be measured in terms of the quantity

$$|\mathbb{E}[\varphi(X_T)] - \mathbb{E}[\varphi(X_T^n)]|,$$

which is called the weak error. Thus, our objective is to investigate the weak error of X_T, where $(X_t)_{0\le t\le T}$ is approximated by the implicit, Crank–Nicholson, resp. explicit Euler scheme. Note, in case of the explicit Euler scheme, due to stability conditions, we have to consider space and time discretization simultaneously.

There exist some works dealing with the strong error (see e.g., Greksch and Kloeden [7], Gyöngy [8, 9], Hausenblas [10], Shardlow [17], and Yoo [23]) but to the author no literature is known dealing with weak convergence. Moreover, as in the finite dimensional case (see e.g. Bally and Talay [2]) the order of convergence in time can be improved to order 1.

The organization is the following. After the introduction, in the second chapter we state the main results. In the third chapter we recall the necessary facts about the Malliavin calculus, which we apply in the fourth chapter to evolution equations. Chapters five and six are concerned with the proof of our main theorem.

2. Formulation of the main result

In the separable Hilbert space $E = L^2(\mathcal{O})$, let $\mathcal{O} \subset \mathbb{R}^d$ be a bounded open domain with regular boundary $\partial\mathcal{O}$, let $X_t(x)$, $t \ge 0$, be the solution of the stochastic partial differential equation

$$\begin{cases} dX_t &= (AX_t + f(X_t))\,dt + Q^{\frac{1}{2}}\,d\omega(t), \quad t \in (0,T], \\ X_0 &= x_0 \in H, \end{cases} \tag{2.1}$$

where A is a second order differential operator with homogeneous boundary condition, Q a linear operator in E, $f : E \to E$ bounded, and $\omega(t)$ is an E-valued cylindrical Wiener process defined on a stochastic basis $(\Omega, \mathcal{F}, \mathbb{P})$ with filtration $\mathcal{F}_t$. The Hilbert–Schmidt norm of an operator is denoted by $\|\cdot\|_{HS}$. In the following we assume

H1. The linear operator A is an infinitesimal generator of an analytic semigroup $S(t)$, $t \ge 0$, in E such that for any $t \in (0,T]$

$$\int_0^t \|S(s)Q^{\frac{1}{2}}\|_{HS}^2\,ds < \infty.$$

H2. There exists an $\alpha \in (0, 1/2)$ such that for every $t > 0$,

$$\int_0^t s^{-2\alpha} \|S(s)Q^{\frac{1}{2}}\|_{HS}^2 \, ds < \infty.$$

H3. Q is a linear self–adjoint nonnegative operator in E which commutes with A.

H4. The function $f : E \to E$ is five times Fréchet differentiable with uniformly bounded and Lipschitz continuous derivatives up to order 5.

One of the most elementary schemes is the implicit Euler scheme, which is, in contrast to the explicit Euler scheme, unconditionally stable. This means that we can consider only time discretization. We denote by τ_n the time step size. Now, using the implicit scheme, we approximate the differential quotient by

$$\frac{X^n_{t+\tau_n} - X^n_t}{\tau_n} \quad \approx \quad AX^n_{t+\tau_n} + f(X^n_t) + Q^{\frac{1}{2}} \, (\omega(t+\tau_n) - \omega(t)) \, .$$

The operator A is approximated by $\tau_n^{-1} \left[(I - \tau_n A)^{-1} - I \right]$, which coincides with the Yosida approximation of A. So, if x^n_k denotes the approximation of $X_{k\tau_n}$, then

$$\left\{ \begin{array}{rcl} x^n_{k+1} & = & (I - \tau_n A)^{-1} x^n_k + \tau_n f(x^n_k) \, + \sqrt{\tau_n} \, Q^{\frac{1}{2}} \, \Delta\xi^k_n, \\ x^0_n & = & P_n x_0, \end{array} \right. \tag{2.2}$$

where $\Delta\xi^k_n$ are independent normal distributed E-valued random variables with mean zero and variance 1.

An application of the trapezoidal rule yields after some minor manipulation another unconditional stable scheme, the Crank–Nicholson scheme. Here, the operator A is approximated by

$$\tau^{-1}(I + \frac{\tau_n}{2} A)(I - \frac{\tau_n}{2} A)^{-1} - I,$$

which results in the scheme

$$\left\{ \begin{array}{rcl} x^n_{k+1} & = & (I + \frac{\tau_n}{2} A)(I - \frac{\tau_n}{2} A)^{-1} x^n_k + \tau_n f(x^n_k) \, + \sqrt{\tau_n} \; Q \, \Delta\xi^k_n, \\ x^0_n & = & P_n x_0. \end{array} \right. \tag{2.3}$$

Now, we can formulate the following theorem:

Theorem 2.1. *Let E be a separable Hilbert space, and $\tilde{E}$ a finite dimensional subspace of E such that Im$(Q) \supset \tilde{E}$. Let $\varphi : \tilde{E} \to \mathbb{R} \in L^\infty(\tilde{E})$ be a measurable function and $X_t(x)$ a solution to Equation* (2.1)*, where A, f and Q satisfy the assumptions* H1,…,H4. *We denote by $X^n_t(x)$ the approximation by the implicit Euler scheme or Crank–Nicholson scheme. Then we have*

$$|\mathbb{E}\left[\varphi(X_t(x))\right] - \mathbb{E}\left[\varphi(X^n_t(x))\right]| \leq C \cdot \tau_n \cdot |\varphi|_\infty.$$

In case of the explicit Euler scheme, we can formulate a similar theorem, but since the scheme is in general not stable, we have to discretize space and time simultaneously. Thus, we denote by the triple (E^n, A_n, P_n) the space discretization corresponding to the time step τ_n, where A_n is a bounded operator on E^n approximating A, the operator P_n denotes a 'projection' from E into E^n, where

E^n is finite dimensional with dimension d_n. Let E_n be an 'interpolation' operator that associates to each element of E^n an element of E. Using the explicit Euler scheme, x_k^n is given by

$$\begin{cases} x_{k+1}^n &= (1+\tau_n A_n)x_k^n + \tau_n P_n f(E_n x_k^n) + \sqrt{\tau_n}\, Q_n\, \Delta\xi_n^k, \\ x_n^0 &= P_n x_0, \end{cases} \tag{2.4}$$

where ξ is a d_n-dimensional standard Gaussian random variable $\mathcal{N}(0, P_n Q E_n)$. Assume additionally that the space discretization satisfies the following assumptions:

(A) (a) $E, E^1, E^2, E^3, \dots$ are all real or all complex valued Banach spaces. All norms will be denoted by $|\cdot|_E$.
(b) P_n is a bounded linear operator, satisfying $|P_n x|_E \le p|x|_E$ for all $n \ge 1$, $x \in E$ and for some $p \ge 0$.
(c) E_n is a bounded linear operator, satisfying $|E_n x|_E \le q|x|_E$ for all $n \ge 1$, $x \in E$ and for some $q \ge 0$, $pq \le 1$.
(d) $P_n E_n x = x$ for all $n \ge 1$ and $x \in E^n$.

Moreover we assume the following stability condition.

(B) A_n is a bounded operator and there exists some $M < \infty$, such that, for $\omega \in \mathbb{R}$,

$$|e^{A_n t}|_E \le M e^{\omega t} \quad \text{for} \quad t \ge 0,\ n \ge 1.$$

Now, we can formulate the following theorem.

Theorem 2.2. *Let (E^n, A_n, P_n), $n \ge 1$, be a finite dimensional approximation of E satisfying* **(A)**, *and* **(B)** *and the stability conditions*

$$\textbf{(C)} \qquad \left|(I+\tau_n A_n)^k\right| \le M e^{\tilde\omega \tau_n k} \quad \textit{for some } k \ge 1 \textit{ and } \tilde\omega \in \mathbb{R}.$$

Then we have under the assumptions of Theorem 2.1

$$\begin{aligned} &|\mathbb{E}\left[\varphi(X_t(x))\right] - \mathbb{E}\left[\varphi(X_t^n(P_n x))\right]| \\ &\qquad \le C\cdot\tau_n|\varphi|_\infty + \left|\mathbb{E}[\varphi(\hat X_t(x))] - \mathbb{E}[\varphi(X_t(x))]\right| \\ &\le\ C\,|\varphi|_\infty \left(\tau_n + \|X_t(x) - \hat X_t(x)\|_2\right), \end{aligned}$$

where $\hat X_t(x)$ denotes the approximation of $X_t(x)$ by space discretization.

Remark 2.3. Note that the constant C in Theorem 2.1 and Theorem 2.2 depends on $\tilde E$, i.e., (see Remark 3.7)

$$C \sim C(\tilde E) = \left(\inf_{v\in\tilde E} \frac{\|S(t)Q^{\frac12}v\|}{\|v\|}\right)^k,$$

for some fixed $k \in \mathbb{N}$. This means, if $E_m \subset E$, $m \in \mathbb{N}$, is a sequence of finite dimensional spaces approximating E, then $C(E_m)$ tends to infinity as $m \to \infty$. Thus, it is important that $\tilde E$ is fixed.

3. Some results of the Malliavin calculus

There exist several approaches of the Malliavin calculus, see e.g., Bismut's [3] approach based on a perturbation argument involving the Girsanov theorem, or the functional analytic approach due to Stroock [19]. Most deal with the case U is finite dimensional and use, in the integration by parts setting, matrices, co-matrices, and determinants. Shigekawa [18] (see also Sugita [20]) also considers the case U finite dimensional, but uses instead of co-matrices the tensor product. Bally and Pardoux [1], Ocone [14] and Pardoux and Zhang [15] generalize the concept to the infinite dimensional case, more precisely to SPDEs driven by space time Wiener processes in one variable, the Brownian sheet. But the requirement of a Hilbert–Schmidt operator for existence of solutions on one side and the nondegeneracy assumptions on the noise on the other side, leads only to finite dimensional results. Nualart and Zakai [13] consider the case of SPDEs driven by higher dimensional space-time white noise. Since the formal solution does not belong to an appropriate space, this leads to generalized functionals. Fuhrmann [6] uses the Malliavin Calculus to show regularity results for the associated semigroup. In this chapter, we recall the definition and some propositions, which are necessary for our computation.

First, let us define the abstract Wiener space. For a fixed $T > 0$, let Ω be the Banach space formed by all continuous functions

$$\omega : [0, T] \to U,$$

such that $\omega(0) = 0$ with norm $\|\omega\| = \sup_{0 \le t \le T} |\omega(t)|_{U_0}$. Let H be the Cameron–Martin subspace of all absolutely continuously differentiable functions with square integrable derivative equipped with the inner product

$$\langle h_1, h_2 \rangle_H = \int_0^T \langle \dot{h}_1(s), Q\dot{h}_2(s) \rangle_U ds = \int_0^T \langle \dot{h}_1(s), \dot{h}_2(s) \rangle_{U_0} ds,$$

where $\dot{h}_i(s) = \frac{d}{ds} h_i(s)$ for $i = 1, 2$ and $U_0 = Q^{-\frac{1}{2}} U$, $Q^{-\frac{1}{2}}$ the pseudoinverse of $Q^{\frac{1}{2}}$. For $l : \Omega \to \mathbb{R} \in \Omega^*$, we denote by $l(\omega)$ the entity $\langle l, \omega \rangle_H$ which equals $\int_0^T \langle \dot{l}(s), d\omega(s) \rangle_{U_0}$. Let P be a zero mean Gaussian measure on Ω determined by

$$\mathbb{E}^P [\exp(i\langle \xi, \omega \rangle_H)] = \int_\Omega \exp(i\langle \xi, \omega \rangle_H) P(d\omega) = \exp(-\frac{1}{2}|\xi|_H),$$

$\xi \in \Omega^*$ and $|\cdot|_H$ is the norm of H induced by the inner product of H. Now, we are acting on the abstract Wiener space (Ω, H, P), with ONB $e_{i,k} = h^i u_k$, $h^1, h^2, \ldots$, ONB in $\{h : [0, T] \to \mathbb{R}\}$ equipped with inner product $\langle h^1, h^2 \rangle = \int_0^T \dot{h}^1(s) \dot{h}^2(s)\, ds$. We use $L^p(\Omega; E)$ to denote all measurable functions $F : \Omega \to E$ such that $|F|_E \in L^p(\Omega) := L^p(\Omega, P)$.

Let E be a separable Hilbert space. Next, we introduce the operators D, δ and $\mathcal{L}$, but first we define the class of E-valued polynomials on Ω. A function $F : \Omega \to E$ is called an E-valued polynomial if it is a linear combination of functions $l(\omega)^m e$, $m \ge 0$, $l \in \Omega^*$ and $e \in E$. Let $\mathcal{P}(E)$ be the space of all functions which can be written as polynomials, i.e., as

$$F(\omega) = \sum_i^m p_i(l_1^i(\omega), l_2^i(\omega), \dots, l_{n_i}^i(\omega))\epsilon_i \tag{3.1}$$

for some polynomial functions $p_i(x_1, \dots, x_{n_i}) : \mathbb{R}^{n_i} \to \mathbb{R}$, for some $l_i^j \in \Omega^*$, $i = 1, \dots, n_i$, and $\epsilon_j \in E$ for $j = 1, \dots, m$ and $m \geq 0$. It can be shown that $\mathcal{P}(E)$ is a dense subspace of $L^p(\Omega; E)$.

Let $\Lambda(H, E)$ be the space of all Hilbert–Schmidt operators $L : H \to E$ equipped with norm $|L|_{\Lambda(H,E)} := \sum_{i,k} \langle Le_{i,k}, Le_{i,k}\rangle_E$. For a function $F : \Omega \to E \in \mathcal{P}(E)$ its derivative $DF : \Omega \to \Lambda(H, E) \in \mathcal{P}(H \otimes E)$ is defined by

$$DF(\omega)[h] := \frac{d}{d\epsilon} F(\omega + \epsilon h)\Big|_{\epsilon=0}, \quad h \in \Omega.$$

A short calculation shows for $F(\omega) = \sum_i^m p_i(l_1^i(\omega), l_2^i(\omega), \dots, l_{n_i}^i(\omega))\varepsilon_i \in E$ with polynomials $p_i(x_1, \dots, x_{n_i}) \in \mathbb{R}$ for $x_j \in \mathbb{R}$, $j = 1, \dots, n_i$, $l_j^i \in \Omega^*$ and $\varepsilon_i \in E$, $i = 1, \dots, m$,

$$DF(\omega) = \sum_i^m \sum_j^{n_i} \frac{dp_i}{dx_j}(l_1^i(\omega), \dots, l_{n_i}^i(\omega))\, l_j^i \otimes \varepsilon_i$$

and

$$DF(\omega)[h] = \sum_i^m \sum_j^{n_i} \frac{dp_i}{dx_j}(l_1^i(\omega), \dots, l_{n_i}^i(\omega))\, \langle l_j^i, h\rangle_H \varepsilon_i.$$

The operator D is closable as an operator from $L^p(\Omega; E)$ to $L^p(\Omega; \Lambda(H, E))$, for any $p \geq 1$. Further, we can iterate the operation D to get $D^k F : \Omega \to \Lambda_k(H, E) \in \mathcal{P}(\Lambda_k(H, E))$, which is defined by

$$D^k F = D \circ D^{k-1} F.$$

Here $\Lambda_k(H, E)$ denotes $\Lambda(H, \Lambda_{k-1}(H, E))$. Let $f : \Omega \to \Lambda(H, E)$. Next, we define the adjoint operator δ from $\mathcal{P}(\Lambda(H, E))$ into $\mathcal{P}(E)$ by the relation

$$\int_\Omega \delta f(\omega) \otimes G(\omega) P(d\omega) = \int_\Omega \sum_{i,k} f(\omega)[e_{i,k}] \otimes DG(\omega)[e_{i,k}] P(d\omega),$$

or

$$\mathbb{E}^P \Big[\delta f \otimes G\Big] = \mathbb{E}^P \Big[\sum_{i,k} f[e_{i,k}] \otimes DG[e_{i,k}]\Big]$$

for $G \in \mathcal{P}(E)$. Integration by parts shows

$$\delta f(\omega) = \operatorname{Tr}(Df(\omega)) - f(\omega)[\omega],$$

where the trace of $A \in \Lambda_2(H;E)$ is defined by $\sum_{i,k} Ae_{i,k}[e_{i,k}] \in E$. To be more explicit, for $f(\omega) = p(l_1(\omega),\dots,l_n(\omega))l \otimes \varepsilon$, $l \in \Omega^*$ and $\varepsilon \in E$,

$$\delta F(\omega) = \left\{ \sum_{k=1}^{n} \frac{dp}{dx_k}(l_1(\omega),\dots,l_n(\omega))\langle l, l_k\rangle_H - p(l_1(\omega),\dots,l_n(\omega))l[\omega] \right\} \varepsilon.$$

Finally, we define the Malliavin covariance operator. Let $F, G : \Omega \to E$. First, we define the bilinear map $\langle F, G\rangle_{\mathcal{L}} : E \to E$ by

$$\langle F, G\rangle_{\mathcal{L}} := \sum_{i,k} DF[e_{i,k}] \otimes DG[e_{i,k}].$$

If F equals G, then $\gamma_F := \langle F, F\rangle_{\mathcal{L}}$ is called the Malliavin covariance operator of F. Another way to define the covariance operator is to introduce first the Ornstein–Uhlenbeck operator $\mathcal{L} : \mathcal{P}(E) \to \mathcal{P}(E)$ by

$$\mathcal{L}F(\omega) := \delta(DF)(\omega) = \mathrm{Tr} D^2 F(\omega) - DF(\omega)[\omega],$$

which is closable and, in case $F = P(l_1[\omega],\dots,l_n[\omega])\epsilon$,

$$\begin{aligned} \mathcal{L}F(\omega) \quad &= \quad \Big\{ \sum_{j,k=1}^{n} \frac{dp}{dx_k dx_j}(l_1(\omega),\dots,l_n(\omega))\langle h_j, h_k\rangle_H \\ &\qquad - \sum_{j=1}^{n} \frac{dp}{dx_j}(l_1(\omega),\dots,l_n(\omega))l_j[\omega] \Big\}\epsilon, \end{aligned}$$

and its semigroup, and then to define the covariance operator by the identity

$$\langle F, G\rangle_{\mathcal{L}} \quad = \quad \mathcal{L}(F \otimes G) - F \otimes \mathcal{L}G - G \otimes \mathcal{L}F.$$

Proposition 3.1. *A short calculation shows the following rules:*

- *Let $\varphi : E \to \mathbb{R}$. Then we have, for $F \in \Lambda(\Omega, E)$,*
$$D\varphi(F)[h] = (\partial\varphi)(F)[DF[h]] = \langle \partial\varphi(F), DF[h]\rangle_E.$$
- *We have for F, $G \in \mathbb{L}^1$, $\delta(FDG) = \langle F, G\rangle_{\mathcal{L}} - F\mathcal{L}G$.*
- *For F, G, and $J \in \Lambda(\Omega, E)$, we have*
$$\langle \langle F, G\rangle_{\mathcal{L}}, J\rangle_{\mathcal{L}} = \langle \langle F, G \otimes J\rangle_{\mathcal{L}}\rangle_{\mathcal{L}} + \langle \langle D^2 G, DF \otimes DJ\rangle_{\mathcal{L}}\rangle_{\mathcal{L}}.$$

Before continuing, it will be useful to have some norms on $L^p(\Omega)$ at our disposal. Let $\Lambda(H, E)$ be the space of all Hilbert–Schmidt operators $L : H \to E$ equipped with norm $|L|_{\Lambda(H,E)} := \sum_{i,k}\langle Le_{i,k}, Le_{i,k}\rangle_E = \mathrm{Tr}(LL^*)$, where $e_{i,k}$, $i,j \in \mathbb{N}$ denotes an (=any) ONB in H. First, we consider the operator D. Its domain is denoted by $\mathbb{L}^{1,p}$, and we define the norm

$$\|F\|_{1,p}^E \quad := \quad \mathbb{E}^P\left[|F|_E^p + |DF|_{\Lambda(H,E)}^p\right]^{\frac{1}{p}}.$$

Further we denote by $\mathbb{L}^{k,p}$ the completion of $\mathcal{P}(E)$ with respect to the norm

$$\|F\|_{k,p}^E \quad := \quad \mathbb{E}^P\left[|F|_E^p + \sum_{i=1}^{k} |D^i F|_{\Lambda_i(H,E)}^p\right]^{\frac{1}{p}}.$$

$\mathbb{D}^{j,\infty}$ will denote the space $\cap_{p\geq 1}\mathbb{D}^{j,p}$, $j\geq 1$. Another way of introducing the norm is by the operator $\mathcal{L}$, i.e.,

$$\|F\|_{s,p}^{E,\mathcal{L}} := \mathbb{E}^P\left[\left|(I-\mathcal{L})^{\frac{s}{2}} F\right|_E\right],$$

but it can be shown that both norms are equivalent (see Sugita [21, Theorem 2.4]). Moreover Meyer [12] obtained the following result.

Proposition 3.2. *(see also Sugita* [21, Theorem 3.1]*) Let $p,q>1$, such that $p^{-1}+q^{-1}=r^{-1}<1$, and let $k=0,1,2,\ldots$. Then for every $F\in\mathbb{D}_{p,k}$, and $G\in\mathbb{D}_{q,k}$, we have $F\otimes G\in\mathbb{D}_{r,k}$, and there exists a positive constant $C_{p,q,k}$ such that*

$$\|F\otimes G\|_{r,k}^E \leq C_{p,q,k}\|F\|_{p,k}^E\|G\|_{q,k}^E.$$

For simplicity, we denote the operator $\langle F,F\rangle_{\mathcal{L}}:E\to E$ by γ_F.

Definition 3.3. *We will say that the random variable F satisfies a nondegeneracy assumption A_p in E, if the operator $\gamma_F=\langle F,F\rangle_{\mathcal{L}}:E\to E$ is a.s. invertible and its inverse $\Gamma_F:=\gamma_F^{-1}$ satisfies*

$$|\Gamma_F|_{L(E,E)}\in L^p(\Omega).$$

Since $\gamma_F\Gamma_F=I$ and from the chain rule, we see that

$$D\Gamma_F = -\Gamma_F^2\, D\gamma_F.$$

Our main ingredient is the following integration by part setting.

Proposition 3.4. *Let $F\in\mathbb{D}^{2,\infty}$ satisfy the nondegeneracy assumption A_p, with $p\geq 3$. Let $G:\Omega\to E\in\mathbb{D}^{1,\infty}$. Let H be defined by*

$$\begin{aligned} H[F,G] &:= \delta(\langle\Gamma_F DF,G\rangle_E)\\ &= -\Big\{ Tr(\Gamma_F\langle F,G\rangle_{\mathcal{L}}) + Tr(\langle\Gamma_F,F\rangle_{\mathcal{L}}G) + \langle\Gamma_F\mathcal{L}F,G\rangle_E\Big\}.\end{aligned}$$

Then, we have for $\varphi\in L^\infty(E;\mathbb{R})$,

$$\mathbb{E}^P\left[Tr((\partial\varphi)(F)\otimes G)\right] = \mathbb{E}^P\left[\varphi(F)\otimes H[F,G]\right]$$

and moreover

$$|\mathbb{E}^P\left[Tr((\partial\varphi)(F)\otimes G)\right]| \leq \||\Gamma_F|^2_{L(E,E)}\|_p\|F\|_{q,2}^{E\ 2}\|G\|_{r,1}^E,$$

for some q and r large enough.

Remark 3.5. Let $k\geq 1$ arbitrary. Let $F\in\mathbb{D}^{2,\infty}$ satisfy the nondegeneracy assumption A_p, with $p\geq 2k+1$. Define $H^k[F,G]:=H^{k-1}[F,H[F,G]]$. Induction implies for $k\geq 1$,

$$|\mathbb{E}^P\left[(\partial^k\varphi)(F)[G_1,\ldots,G_k]\right]| \leq \|\Gamma_F^2\|_p\|F\|_{q,q'}^{q''}\|G_1\|_{r_1,r_1'}\cdots\|G_k\|_{r_k,r_k'},$$

for some q, q', q'', r_i and r_i', $i=1,\ldots,k$ large enough.

Proof. Let $\varphi \in C^1(E)$. A short calculation shows that we have

$$\langle \varphi(F), F\rangle_{\mathcal{L}} = \sum \langle DF[e_i], \partial\varphi(F)\rangle_E \otimes DF[e_i],$$

and by the definition of the tensor product,

$$\begin{aligned}\langle \varphi(F), F\rangle_{\mathcal{L}} &= \sum \left(DF[e_i] \otimes DF[e_i]\right)\left(\partial\varphi(F)\right)\\ &= \gamma_F\left(\partial\varphi(F)\right).\end{aligned}$$

On the other hand, we have by the linearity of D,

$$\begin{aligned}&\mathbb{E}^P\left[\left\langle \Gamma_F \langle \varphi(F), F\rangle_{\mathcal{L}}, G\right\rangle_E\right]\\ &\qquad = \mathbb{E}^P\left[\left\langle \Gamma_F\left(\sum_i D\varphi(F)[e_i] \otimes DF[e_i]\right), G\right\rangle_E\right],\end{aligned}$$

and, since $D\varphi(F)[e_i] \in \mathbb{R}$,

$$= \mathbb{E}^P\left[\sum_i D\varphi(F)[e_i] \otimes \langle \Gamma_F DF[e_i], G\rangle_E\right].$$

As δ is the adjoint of D, we can continue;

$$\mathbb{E}^P\left[\langle \Gamma_F\langle \varphi(F), F\rangle_{\mathcal{L}}, G\rangle_E\right] = \mathbb{E}^P\left[\varphi(F) \otimes \delta\left(\langle \Gamma_F DF, G\rangle_E\right)\right]$$

and we get

$$\begin{aligned}\mathbb{E}^P\left[(\partial\varphi(F)[G])\right] &= \mathbb{E}^P\left[\langle \partial\varphi(F), G\rangle_E\right]\\ &= \mathbb{E}^P\left[\left\langle \Gamma_F\langle \varphi(F), F\rangle_{\mathcal{L}}, G\right\rangle_E\right]\\ &= \mathbb{E}\left[\varphi(F) \cdot H[F,G]\right].\\ &\leq |\varphi|_\infty \|\Gamma_F\|_3^2 \|F\|_{2,p} \|G\|_{1,q},\end{aligned} \tag{3.2}$$

such that $1/p + 1/q \leq 2/3$. Let φ be only bounded and measurable. Then, there exists a sequence of functions $\varphi_\epsilon \in C^1(E)$ such that φ_ϵ tends to φ in E as ϵ tends to zero. Since the right-hand side of the inequality (3.2) does not depend on $\partial\varphi$, the inequality holds uniform in ϵ and we can pass to the limit. □

3.1. The Malliavin calculus applied to SPDEs

Our points of interest are functionals of the Wiener space, which arise as a solution to a stochastic partial differential equation, i.e., let $X_t : \Omega \to E$ be a solution to (2.1), satisfying the assumptions H1, ..., H4. It follows that X_t belongs to the domain of the Malliavin derivative, and the resulting process satisfies again a stochastic partial differential equation.

To be precise, we identify first U with E and let $DX_t : [0,T] \times \Omega \to E$ be defined by

$$DX_t(\omega)[h] := \frac{d}{d\epsilon} X_t(\omega + \epsilon h)\Big|_{\epsilon=0} = \langle DX_t(\omega), h\rangle_H.$$

Then $DX_t[h]$ satisfies

$$\begin{cases} DX_t[h] = \int_0^t (A + \partial f(X_r)) DX_r[h]\, dr + \int_0^t Q^{\frac{1}{2}}\, \dot{h}(r)\, dr, \\ DX_0[h] = 0. \end{cases} \tag{3.3}$$

Equation (3.3) is well defined since $\partial f(x)$, $x \in E$, is Lipschitz continuous and uniformly bounded in x. Next, to get another expression of DX_t, we define the Jacobian operator J_t of the mapping $x \mapsto X_t(x, \omega)$, where $J_t(x) = \partial_x X_t(x, \omega)$. Since f is smooth enough, $X_t(x)$ has mean square directional derivative $\partial_x X(x)[v]$ at x and in the direction of $v \in E$. It can be shown (see [4, Theorem 5.4.1]) that $J_t(x) = \partial_x X(x)$ is the unique solution of the equation

$$\begin{cases} dJ_t(x)[v] &= (A + \partial f(X_t(x)))\, J_t(x)[v] dt, \\ J_0(x)[v] &= v, \end{cases} \tag{3.4}$$

belonging to $C([0,T]; L^p(\Omega; E))$. Since ∂f is a linear operator, there exists a constant $C = C_{T,p}$, independent of h, such that $\mathbb{E}\left[\int_0^t |J_s(x)[v]|^p\right] \le C\, |v|^p < \infty$. By the method of variation of constants we get another expression for DX_t and for $\gamma_t := \gamma_{X_t}$, i.e.,

$$DX_t[h] \quad = \quad \int_0^t J_{t-s}(X_s) Q^{\frac{1}{2}}\, dh(s) \tag{3.5}$$

and

$$\gamma_t \quad = \quad \int_0^t J_{t-s}(X_s) Q^{\frac{1}{2}} e_i \, \otimes J_{t-s}(X_s) Q^{\frac{1}{2}} e_i \, ds, \tag{3.6}$$

where $e_1, e_2, \dots$ is an ONB in E. Now, we can formulate the following proposition.

Proposition 3.6. *Let $\tilde{E}$ be a finite dimensional subspace of E, $\tilde{P}$ the projection $\tilde{P}: E \to \tilde{E}$. Under the hypothesis of Theorem 2.1, $\tilde{P}X_t : \Omega \to \tilde{E}$ is $(A)_p$ nondegenerate for $p \ge 2$ arbitrary and t strictly larger than zero.*

Proof. By Definition 3.3 we have to investigate γ_t^{-1}. First we compute $\langle \gamma_t u, u\rangle$ for $u \in \tilde{E}$:

$$\begin{aligned} \langle \gamma_t u, u\rangle &= \sum_i \int_0^t \langle J_{t-s}(X_s) Q^{\frac{1}{2}} e_i, u\rangle \langle J_{t-s}(X_s) Q^{\frac{1}{2}} e_i, u\rangle \, ds \\ &= \int_0^t \langle J_{t-s}(X_s) Q^{\frac{1}{2}} u, J_{t-s}(X_s) Q^{\frac{1}{2}} u\rangle \, ds. \end{aligned}$$

Let us define

$$\tilde{\zeta}_t = \inf_{v \in \tilde{E}} \frac{\|S(t) Q^{\frac{1}{2}} v\|}{\|v\|},$$

where t is strictly larger than zero. Note, since $\operatorname{Im}(Q) \supset \tilde{E}$, the entity $\tilde{\zeta}_t$ is bounded away from zero. Moreover, hence $J_t(x)$ coincides with the semigroup $S(t)$ of A, perturbed by the linear and bounded operator $g_s = \partial f(X_s(x))$, i.e.,

$$J_t(x)[u] = S(t)u + \int_0^t S(t-s)\, \partial f(X_s(x)) J(X_s(x))[u]\, dt,$$

where $\partial f(y) : E \to E$ is Lipschitz and uniformly bounded in y. A short calculation shows that for $t > 0$ fixed

$$\tilde{\zeta}_t \sim \inf_{\substack{v \in \tilde{E} \\ y \in E}} \frac{\|J_t(y) Q^{\frac{1}{2}} v\|}{\|v\|} > 0,$$

and

$$\langle \gamma_t u, u \rangle \geq t\, \tilde{\zeta}_T\, \|u\|.$$

The Lax–Milgram Theorem yields the assertion. □

Remark 3.7. Consider a sequence of space approximations E_m of E and define

$$\tilde{\zeta}_{m,t} = \inf_{v \in E_m} \frac{\|S(t) Q^{\frac{1}{2}} v\|}{\|v\|}.$$

Note that a short calculation shows that $\tilde{\zeta}_{m,t}$ tends to zero as m tends to infinity for $t > 0$.

In view of 3.4, it is clear that estimates on the transition probability as a function of the forward variable follow immediately from bounds on $\|\Gamma_F\|_p$, $1 \leq p < \infty$. But, the solution of (2.1) can be also seen as a function of the initial condition $x \in E$, i.e., the backward variable. Now, by means of the Jacobi operator $J(x)$, the integration by parts setting also yields estimates on the transition probability as a function of the backward variable. To be precise, let $X_t(x)$ be the solution of (2.1) with initial condition $X_0 = x \in E$. Now we have

$$\begin{aligned} \partial_x \mathbb{E}\left[\varphi(\tilde{P} X_t(x))\right] &= \mathbb{E}\left[(\partial\varphi)(\tilde{P} X_t(x)) \tilde{J}_t(x)\right] \qquad (3.7)\\ &= \mathbb{E}\left[\varphi(\tilde{P} X_t) H[\tilde{P} X_t, \tilde{J}_t(x)]\right], \end{aligned}$$

where $\tilde{J}_t(x) = \frac{\partial}{\partial x} \tilde{P} X_t(x)$. By the Clarc–Ocone formula, we can see that the formula is related to the Bismut–Elworthy formula (see DaPrato and Zabczyk [5, Lemma 7.1.3]). The difference is that they used the chain rule implicitly by identifying $\mathbb{E}[J_t(\varphi, x)]$ with $\mathbb{E}^x[J^{\varphi}_{t-s}(X_s(x)) J_s(x)]$, where $J_t(\varphi, x) = \frac{\partial}{\partial x}\mathbb{E}[\varphi(X_t(x))]$. Let $\tilde{J}^i_t(x)$ be defined by $\tilde{J}^i_t(x) = \frac{\partial^i}{\partial x^i} \tilde{P} X_t(x)$. Iteration of equation (3.7) yields

$$\begin{aligned} \partial^x_\alpha \mathbb{E}[\varphi(\tilde{P} X_t(x))] &= \sum_{\nu=1}^{\alpha} \mathbb{E}\left[(\partial^y_\nu \varphi)(y)\big|_{y=\tilde{P} X_t(x)} \sum J_t^{|I_1|}(x) \cdots J_t^{|I_\nu|}(x)\right] \\ &=: \sum_{\nu=1}^{\alpha} \mathbb{E}\left[(\partial^y_\nu \varphi)(y)\big|_{y=\tilde{P} X_t(x)} \psi_\nu(t,x)\right], \end{aligned}$$

where the second sum runs over all partitions $\{I_1, \ldots, I_\nu\}$ of the set $\{1, \ldots, \alpha\}$ with length ν. Note, if $X_t(x) \in \mathbb{D}^{k,\infty}$, it follows that $D^l J^i_t(x)$ is well defined for $i + l \leq k$

and $J_t^i(x) \in \mathbb{L}^{l,\infty}$. Assume $X_t(x) \in \mathbb{L}^{k+1,\infty}$ and $\tilde{P}X_t(x)$ is $(A)_{2k+1}$ nondegenerate. Now, it follows from Proposition 3.4, that for $|\alpha| \le k$, $\tilde{\Gamma}_t = (\gamma_t|_{\tilde{E}\times\tilde{E}})^{-1}$,

$$\begin{aligned}
\partial_\alpha^x \mathbb{E}[\varphi(\tilde{P}X_t(x))] &= \sum_{\nu=1}^{\alpha} \mathbb{E}[\varphi(\tilde{P}X_t(x)) H_\nu(\tilde{P}X_t(x), \psi_\nu(t,x))] \\
&\le |\varphi|_\infty \|\tilde{\Gamma}_t\|_{2k+1}^{2k} \|\sum_{\nu=1}^{\alpha} \tilde{\psi}_\nu(t,x)\|_{k,r} \|\tilde{P}X_t\|_{k+1,q} \\
&\le C(t)\cdot|\varphi|_\infty \cdot \frac{1+|x|^q}{t^m}, \qquad (3.8)
\end{aligned}$$

for some $m, r, q < \infty$.

3.2. The Malliavin calculus and the approximation

Our next point of interest is the Malliavin derivative of the approximation. To cover all cases we define the operator $F_\tau(A)$; in case of

$$\begin{cases}
\text{the } \textit{explicit Euler scheme}, & F_\tau(A) = (I+\tau A), \\
\text{the } \textit{implicit Euler scheme}, & F_\tau(A) = (I-\tau A)^{-1}, \quad \text{and} \\
\text{the } \textit{Crank–Nicholson scheme}, & F_\tau(A) = (I+\frac{1}{2}\tau A)(I-\frac{1}{2}\tau A)^{-1}.
\end{cases} \qquad (3.9)$$

The operator itself is approximated by $A_\tau^{app} := \tau^{-1}(F_\tau(A) - I)$. Moreover, considering the implicit Euler scheme or the Crank–Nicholson scheme, let $E^n := E$, $A_n := A$ and $P_n = E_n := I$, and elsewhere assume the space discretization satisfies conditions **(A)**, **(B)** and **(C)**. Note, in case of the implicit Euler scheme or the Crank–Nicholson scheme the stability condition **(C)** is automatically satisfied (see Pazy [16, Theorem 1.5.3]).

Remark 3.8. We have for the difference $A_\tau^{app} - A$ in case of

$$\begin{cases}
\text{the explicit Euler scheme}, & 0, \\
\text{the implicit Euler scheme}, & \tau A^2(I-\tau A)^{-1}, \quad \text{and} \\
\text{the Crank–Nicholson scheme}, & \tau A^2(I-\frac{1}{2}\tau A)^{-1}.
\end{cases}$$

Let $f_n(x) := P_n f(E_n x)$, and $Q_n := P_n Q E_n$. Moreover $X_t^n(x)$ denotes the approximation of $X_t(x)$ starting at $P_n x$. If it is not necessary, we omit the initial point x. Now, we have for $k\tau_n < t \le \tau_n(k+1)$, $k \ge 1$,

$$\begin{cases}
DX_t^n[h] = DX_{k\tau_n}^n[h] + (A_{\tau_n}^{app} + \partial f_n(X_{k\tau_n}^n)) DX_{k\tau_n}^n[h]\left(t - \frac{k}{n}\right) \\
\qquad\qquad + \int_{k\tau_n}^t Q_n^{\frac{1}{2}} \dot{h}(r)\,dr, \\
DX_0^n[h] = 0,
\end{cases}$$

and

$$\begin{cases}
J_t^n(x)[u] = J_{k\tau_n}^n[u] + (A_{\tau_n}^{app} + \partial f_n(X_{k\tau_n}^n(x)) J_{k\tau_n}^n(x)[u]\,(t-k\tau_n) \\
\qquad\qquad + Q_n^{\frac{1}{2}}\left(\omega(t) - \omega(k\tau_n)\right), \\
J_0^n(x)[u] = u.
\end{cases}$$

By induction we get

$$J^n_{k\tau_n} P_n(x)[u] = F_{\tau_n}(A_n)^k P_n u + \tau_n{}^2 \sum_{i=0}^{k-1} F_{\tau_n}(A_n)^i \partial f_n(X^n_{k\tau_n}(x)) P_n u$$
$$+ \sum_{i=0}^{k-1} F_{\tau_n}(A_n)^i Q_n P_n u \left(\omega((i+1)\tau_n) - \omega(i\tau_n)\right)$$

for $u \in E$. Let $T = m\tau_n$. To show

$$\mathbb{E}\left[\sup_{0\le k\le N} \left|J^n_{k\tau_n} P_n(x)[u]\right|_E^p\right] \le C\,|u|_E^p < \infty,$$

we have to mimic a discrete version of the proof [5, Theorem 5.4.1] replacing all integrals by sums. In analogy to (3.5) and (3.6), we have for $k\tau_n < t \le (k+1)\tau_n$,

$$DX^n_t[h] = \sum_{i=0}^{k} J^n_{(k-i)\tau_n}(X^n_{(i-1)\tau_n}) \int_{i\tau_n}^{(i+1)\tau_n \wedge t} Q^{\frac{1}{2}} \dot{h}(r)dr$$

and

$$\gamma^n_t = \tau_n \sum_{i=0}^{k-1} \sum_j J^n_{(k-i)\tau_n}(X'^n_{(i-1)\tau_n})) Q_n^{\frac{1}{2}} e_j \otimes J^n_{(k-i)\tau_n}(X'^n_{(i-1)\tau_n}) Q_n^{\frac{1}{2}} e_j.$$

As done above, we can formulate the following proposition.

Proposition 3.9. *Let $(X_t)_{t\ge 0}$ be a solution to (2.1) and $(X^n_t)_{t\ge 0}$ its approximation given by (2.2), satisfying the assumptions of Theorem 2.1, resp. Theorem 2.2. Then $\tilde{P}X^n_t$ is $(A)_p$ nondegenerate in $\tilde{E}$ for $p \ge 2$ arbitrary and t strictly larger than zero.*

Proof. The proof is straightforward in analogy to the proof of Proposition 3.6. □

4. Proof of the main result

Since the proof of the preceding theorem is based upon the following technical lemma, we state it here. The proof is given in the next section.

Lemma 4.1. *Let $(X_t)_{t\ge 1}$ be a solution of (2.1) in E, where A, f and Q satisfies the assumptions* H1,...,H4. *Let $\tilde{E}$ be a finite dimensional subspace of E. Let the function u be defined by $u(x,t) := \mathbb{E}^x[\varphi(X_{T-t})]$, where $\varphi : \tilde{E} \to \mathbb{R}$ is bounded and measurable. Then for any $\alpha < \alpha_0 = 5$ and for any functions $g_i : E \to E$, $i = 1,\dots,\alpha$ of polynominal growth, there exists a nondecreasing function $K(T)$ and positive constants q and m, uniform with respect to n and T, such that we have, uniform for all $t \in [0,T]$,*

$$|\mathbb{E}\left[\partial_\alpha u(t, X_t(x))[g_1(X_t(x)),\dots,g_\alpha(X_t(x))]\right]| \le K(T)|\varphi|_\infty \frac{1+|x|^q}{T^m}$$

and for $g_i : E \times E \to E$ of polynominal growth,

$$\left|\mathbb{E}\left[\partial_\alpha u(t,\hat{X}_t^n(x))[g_1(\hat{X}_t^n(x),X_t(x)),\ldots,g_\alpha(\hat{X}_t^n(x),X_t(x))]\right]\right| \leq K(T)|\varphi|_\infty \frac{1+|x|^q}{T^m}.$$

The weak error can be written as

$$\begin{aligned} Err(\varphi,T,x;n) &:= |\mathbb{E}[\varphi(X_T)] - \mathbb{E}[\varphi(\hat{X}_T^n)]| + |\mathbb{E}[\varphi(\hat{X}_T^n)] - \mathbb{E}[\varphi(X_T^n)]| \\ &:= I + II \end{aligned}$$

where $\hat{X}_T^n$ denotes the approximation of X_T only in space. Note that the first difference is zero in case of the explicit or Crank–Nicholson scheme. In case of the explicit scheme, I equals zero if $\tilde{E} \subset E_n$. Assume $\tilde{E} \not\subset E_n$. Then the Taylor formula yields

$$\begin{aligned} I &\leq |\mathbb{E}\left[\partial\varphi(X_T)\left(X_T - \hat{X}_T^n\right)\right] \\ &\leq |\varphi|_\infty \|X_T - \hat{X}_T^n\|_2. \end{aligned}$$

To handle the second difference II, we follow Talay and Tubaro [22] and define the functional $u(x,t) := \mathbb{E}^x[\varphi(\hat{X}_{T-t}^n)] = \mathbb{E}^x[\varphi(\tilde{P}\hat{X}_{T-t}^n)]$. Assume $u(t,x)$ satisfies the Kolmogorov equation for φ, i.e.,

$$\begin{cases} \partial_t u = Lu & \text{for } t \geq 0, \\ u(0,x) = \varphi(x) & \text{for } x \in E, \end{cases}$$

where L is the infinitesimal generator of X_t and

$$L\psi(x) = \frac{1}{2}\mathrm{Tr}(Q^{\frac{1}{2}}\partial^2\psi(x)Q^{\frac{1}{2}}) + \langle Ax + f(x), \partial\psi(x)\rangle_E,$$

where $\psi : \tilde{E} \to \mathbb{R}$. Now, the second summand can be written as

$$\begin{aligned} II &= |\mathbb{E}[\varphi(\hat{X}_T)] - \mathbb{E}[\varphi(\hat{X}_T^n)]| \\ &= |\mathbb{E}\left[u(T,x_0) - u(0,X_T^n)\right]|. \end{aligned}$$

Writing the right-hand side as a sum and applying the Itô formula twice, we get

$$\begin{aligned} Err(\varphi,T,x;n) &= \left|\sum_{k=1}^{[Tn]-1} \mathbb{E}\left[u((k+1)\tau_n, \hat{X}_{(k+1)\tau_n}^n) - u(k\tau_n, X_{k\tau_n}^n)\right]\right| \\ &= \sum_{k=1}^{[Tn]} \int_{k\tau_n}^{(k+1)\tau_n} \int_{k\tau_n}^{t} \Big| \mathbb{E}[LLu(s,X_s^n) - LL_z u(s,X_s^n) \\ &\quad - L_z L u(s,X_s^n) + L_z L_z u(s,X_s^n)]\Big|_{z=X_{k\tau_n}^n} \Big| \, ds\, dt \\ &\quad + \sum_{k=1}^{[Tn]} \left| \int_{k\tau_n}^{(k+1)\tau_n} (L_z^{app} - L_z)\, u(s,X_s^n)\Big|_{z=X_{k\tau_n}^n} ds \right|, \end{aligned} \tag{4.1}$$

where for fixed $z \in E$, L_z is defined by

$$L_z\psi(x) = \frac{1}{2}\mathrm{Tr}(Q^{\frac{1}{2}}\partial^2\psi(x)Q^{\frac{1}{2}}) + \langle Az + F(z), \partial\psi(x)\rangle_E,$$

and

$$L_z^{app}\psi(x) = \frac{1}{2}\mathrm{Tr}(Q^{\frac{1}{2}}\partial^2\psi(x)Q^{\frac{1}{2}}) + \langle A^{app}z + F(z), \partial\psi(x)\rangle_E.$$

Setting

$$\Psi(t,x) = \frac{1}{2}\left(L^2u(t,x) - 2L_zLu(t,x) + L^2u(t,x)\right)\big|_{z=x}, \tag{4.2}$$

we first show that

$$\int_0^T |\mathbb{E}[\Psi(t,X_t^n)]|\,dt < \infty.$$

But analyzing term by term, we can see that $\Psi(t,X_t^n)$ is of the form

$$\Psi(t,X_t^n) = \sum_{\alpha\le 4}\partial^4u(t,X_t^n)[g_1(\hat{X}_t^n,X_t^n),\ldots,g_4(\hat{X}_t^n,X_t^n)], \tag{4.3}$$

where $g_1,\ldots,g_4$ are at least of polynomial growth. Thanks to Lemma 4.1, its expectation is finite uniformly in n and t. Next, we investigate the last summand, i.e.,

$$\sum_{k=1}^{[Tn]}\left|\int_{k\tau_n}^{(k+1)\tau_n}(L_z^{app}-L_z)\,u(s,X_s^n)\Big|_{z=X_{k\tau_n}^n}ds\right|.$$

But we have

$$(L_z^{app}-L_z)\,\psi = \begin{cases} \text{for the explicit Euler scheme,} & 0,\\ \text{for the implicit Euler scheme,} & \langle\tau A^2(I-\tau A)^{-1},\psi(x)\rangle_E, \quad \text{and}\\ \text{for the Crank–Nicholson scheme,} & \langle\tau A^2(I-\frac{1}{2}\tau A)^{-1},\psi(x)\rangle_E.\end{cases}$$

Analyzing the difference we can see that $(L_z^{app}-L_z)\,u(t,X_t^n)$ is of the form (4.3). Thus, thanks to Lemma 4.1, its expectation is bounded by $C\tau_n$ uniform t, therefore

$$\sum_{k=1}^{[Tn]}\int_{k\tau_n}^{(k+1)\tau_n}\left|(L_z^{app}-L_z)\,u(s,X_s^n)\Big|_{z=X_{k\tau_n}^n}\right|ds < C\tau_n. \qquad \square$$

5. Proof of Lemma 4.1

Following Bally and Talay [2] we distinguish between small and large t. If t is small, resp. $T-t$ large, we get rid of the derivatives of $u(t,x)$ applying a Bismut–Elworthy-type formula, resp. formula (3.8). Hence the formula fails for $t \to T$; we use Malliavin's integration by parts formula to give an estimate for large t, resp. for small $T-t$.

The case $0 \le t \le \frac{T}{2}$**:** We have to show that we have

$$|\mathbb{E}\left[\partial^\alpha u(t, X_t)\right]| < \infty,$$

for $0 \le \alpha \le 4$. But, by hypothesis H4, i.e., f is Lipschitz continuous and has bounded derivatives up to order 5, we know $X_t \in \cap_{p\ge 2}\mathbb{D}^{5,p}$ and $J_t^i(x) \in \cap_{p\ge 2}\mathbb{D}^{l,p}$, where $i+l \le 5$. Moreover by Proposition 3.6, we know $\tilde{P}X_t$ is $(\mathrm{A})_p$ nondegenerate over $\tilde{E}$ for $p \ge 2$. Applying formula (3.8) we get

$$|\partial^\alpha u(t,x)|_{L^\alpha(E,E)} \le C \cdot |\varphi|_\infty \cdot \frac{1+|x|^q}{(T-t)^m}.$$

Thus, we have

$$\begin{aligned} |\mathbb{E}\left[\partial^\alpha u(t, X_t)[g_1(X_t), \ldots, g_\alpha(X_t)]\right]| &\le |\varphi|_\infty \cdot \frac{C}{(T-t)^m} \mathbb{E}\left[(1+|X_t|_E^q)\right] \\ &\le |\varphi|_\infty \cdot C \cdot \frac{1+|x|_E^q}{(T-t)^m} < \infty \end{aligned}$$

for some $m \in \mathbb{N}$.

The case $\frac{T}{2} \le t \le T$**:** First, we know by Proposition 3.6, resp. Proposition 3.9, that $\tilde{P}X_t$, resp. $\tilde{P}X_t^n$ is nondegenerate for t strictly larger than zero. Moreover, since f is up to order 5 differentiable with bounded and Lipschitz continuous derivatives (see H5), X_t and X_t^n belong to $\cap_{p\ge 2}\mathbb{D}^{5,p}$, and therefore $g(X_t^n, X_t)$ belongs also to $\cap_{p\ge 2}\mathbb{D}^{5,p}$. Thus by Remark 3.5, we can conclude for $\alpha \le 4$ and $\tilde{\Gamma}_t^n = (\gamma_t^n|_{\tilde{E}\times\tilde{E}})^{-1}$,

$$\begin{aligned} &|\mathbb{E}[|\partial^\alpha u(t, X_t^n)[g_1(X_t^n, X_t), \ldots, g_\alpha(X_t^n, X_t)]|_E\,]| \\ &\qquad \le \|\Gamma_t^n\|_m^{2\alpha} \|\hat{X}_t^n\|_{k+1,q} \|g(X_t^n, X_t)_t\|_{k,p}. \\ &\qquad \le C(t) \cdot \frac{1+|x|_E^p}{t^m}. \end{aligned}$$

□

References

[1] V. Bally and E. Pardoux. Malliavin calculus for white noise driven parabolic SPDEs. *Potential Anal.*, 9:27–64, 1998.

[2] V. Bally and D. Talay. The law of the Euler schema for stochastic differential equations: I. *Probab. Theory Relat. Fields*, 104:43–60, 1996.

[3] J. Bismut. Martingales, the Malliavin calculus and hypoellipticity under general Hörmander's conditions. *Z. Wahrscheinlichkeitstheor. Verw. Geb.*, 56:469–505, 1981.

[4] G. Da Prato and J. Zabczyk. *Stochastic equations in infinite dimensions*, volume 44 of *Encyclopedia of Mathematics and Its Applications.* Cambridge University Press, 1992.

[5] G. Da Prato and J. Zabczyk. *Ergodicity for infinite dimensional systems*, volume 229 of *London Mathematical Society Lecture Note Series.* Cambridge University Press, 1996.

[6] M. Fuhrman. Smoothing properties of nonlinear stochastic equations in Hilbert spaces. *NoDEA, Nonlinear Differ. Equ. Appl.*, 3:445–464, 1996.

[7] W. Greksch and P. Kloeden. Time-discretised Galerkin approximations of parabolic stochastic PDEs. *Bull. Aust. Math. Soc.*, 54:79–85, 1996.

[8] I. Gyöngy. Lattice approximations for stochastic quasi-linear parabolic partial differential equations driven by space-time white noise: I. *Potential Anal.*, 9:1–25, 1998.

[9] I. Gyöngy. Lattice approximations for stochastic quasi-linear parabolic partial differential equations driven by space-time white noise: II. *Potential Anal.*, 11:1–37, 1999.

[10] E. Hausenblas. Approximation for semilinear stochastic evolution equations. *Potential Analysis*, 18(2):141–186, 2003.

[11] V.G. Mavrantzas and H.C. Öttinger. Thermodynamically founded atomistic monte carlo for nonequilibrium systems. submitted to *Physical Review Letters*, 2002.

[12] P.A. Meyer. Quelques resultats analytiques sur le semi-groupe d'Ornstein–Uhlenbeck en dimension infinie. In *Theory and application of random fields*, volume 49 of *Lect. Notes Control Inf. Sci.*, pages 201–214. Springer Verlag, 1983.

[13] D. Nualart and M. Zakai. Generalized Brownian functionals and the solution to a stochastic partial differential equation. *J. Funct. Anal.*, 84:279–296, 1989.

[14] D. Ocone. Stochastic calculus of variations for stochastic partial differential equations. *J. Funct. Anal.*, 79(2):288–331, 1988.

[15] E. Pardoux and T. Zhang. Absolute continuity of the law of the solution of a parabolic SPDE. *J. Funct. Anal.*, 112:447–458, 1993.

[16] A. Pazy. *Semigroups of linear operators and applications to partial differential equations*, volume 44. New York etc.: Springer-Verlag, 1983.

[17] T. Shardlow. Numerical methods for stochastic parabolic PDEs. *Numer. Funct. Anal. Optimization*, 20:121–145, 1999.

[18] I. Shigekawa. Derivatives of Wiener functionals and absolute continuity of induced measures. *J. Math. Kyoto Univ.*, 20:263–289, 1980.

[19] D. W. Stroock. The Malliavin calculus, a functional analytic approach. *J. Funct. Anal.*, 44:212–257, 1981.

[20] H. Sugita. On a characterization of the Sobolev spaces over an abstract Wiener space. *J. Math. Kyoto Univ.*, 25:717–725, 1985.

[21] H. Sugita. Sobolev spaces of Wiener functionals and Malliavin's calculus. *J. Math. Kyoto Univ.*, 25:31–48, 1985.

[22] D. Talay and L. Tubaro. Expansion of the global error for numerical schemes solving stochastic differential equations. *Stochastic Ananlysis and its Applications*, 8:483–509, 1990.

[23] H. Yoo. Semi-discretization of stochastic partial differential equations on r^1 by a finite-difference method. *Math. Comput.*, 69:653–666, 2000.

Institut für Mathematik
Universität Salzburg
Hellbrunner Str. 34
5026 Salzburg
Austria

Progress in Probability, Vol. 53, 129–149

Exponential Type Decrease of the Density for Jump Processes with Singular Lévy Measures in Small Time

Yasushi Ishikawa

Abstract. In this paper we consider the asymptotic behaviour of the transition density for processes of jump type as the time parameter t tends to 0. We use Picard's duality method, which allows us to obtain the lower and upper bounds of the density even for the case where the support of Lévy measure is singular. The main result is that, under certain restrictions, the density may exhibit an exponential type decrease as $t \to 0$ according to the accessibility of the objective points by a certain Markov chain.

Introduction

Let $x_t(x)$ be a d-dimensional jump process $x_t(x)$ given by the stochastic differential equation

$$x_t(x) = x + \int_0^t b(x_s(x))ds + \sum_{s \leq t}^{c} \gamma(x_{s-}(x), \Delta z(s)), \tag{0.1}$$

where $z(s)$ denotes an $\mathbf{R}^d$-valued Lévy process of jump type, $\gamma(x, \zeta)$ is a function from $\mathbf{R}^d \times \mathbf{R}^d$ to $\mathbf{R}^d$ and $\sum_{s \leq t}^{c}$ denotes the compensated sum (cf. Section 1).

J. Picard [21] has developed a calculus for showing the existence and the regularity of the transition density $p_t(x, y)$ for $x_t(x)$ (as the density of the law of the process $x_t(x)$), which is applicable to the case where the Lévy measure of the driving process $z(s)$ is not necessarily smooth. This uses a method of adding and subtracting Dirac masses on the Poisson space-time state space (called the *duality method*), instead of previous methods using perturbations in space or time directions developed e.g., by [1], [2] , [3] and [24]. This method, in particular, has the advantage that it can be applied to the case of singular Lévy measures (e.g., it is countably supported).

In this note, we will study the asymptotic behavior of $p_t(x, y)$, for given x and y, as $t \to 0$ under the condition that $z(s)$ has a singular Lévy measure. In particular, its support may be discrete (μ is the sum of point masses). We shall extend the type of processes treated in [22], [23], and apply the method of Picard.

The present study may also be viewed as a continuation and an improvement of the same subject in [9], [10], [11] and [12]. Indeed it will be seen, as in those previous results, that the geometrical quantity $\alpha(x,y)$ attached to the support of the Lévy measure plays an important role in estimating $p_t(x,y)$ for accessible points y (see Section 1 for the definition). Namely if y is accessible, then we have

$$p_t(x,y) \asymp t^{\alpha(x,y)-d/\beta} \quad \text{as} \quad t \to 0. \tag{0.2}$$

If y is not accessible, then the density does not decay in polynomial order of t as $t \to 0$, and makes an exponential type decrease while preserving the positivity (see Theorem 1.4 (b), (c)). The resulting formula in this exponential case is close to what Picard obtained in [22] in the case $d = 1$ and $x_t(x)$ is a Lévy process.

We compose this note as follows. In Section 1 we state our results under necessary assumptions. In Section 2, we give fundamental lemmas (Lemmas 2.1, 2.2). In Sections 3, 4 we give the proof of Theorem 1.4. We prove that the density makes an exponential type decrease in t as $t \to 0$ if y is an asymptotically accessible (but not accessible) point. The proof of the upper bound is the same as [13] Section 4. To show the lower bound, we construct the trajectory concretely and show that the set of such trajectories has positive probability. This subject is much related with the problem of finding "support" of the law on the path space for $t \mapsto x_t(x)$. In this direction, the author owes much to Simon [26].

Lastly, we mention the point of view of Carlen–Pardoux [5], Denis [4], who obtained a criterion for the existence of the density for the solutions of SDE's driven by Poisson random measure of certain type. In order to treat the case where the mean measure of the Poisson random measure is countably supported, they use the theory of Dirichlet forms.

1. Statement of Results

Let $z(s)$ be a Lévy process , $\mathbf{R}^d$-valued, with Lévy measure $\mu(d\zeta)$: $\int_{\mathbf{R}^d}(|\zeta|^2 \wedge 1)\mu(d\zeta) < +\infty$. That is, the characteristic function ψ_t is given by

$$\psi_t(\xi) = E[e^{i\xi.z(t)}] = \exp(it\xi.c + t\int(e^{i\xi.\zeta} - 1 - i\xi.\zeta 1_{\{|\zeta|\le 1\}})\mu(d\zeta)).$$

We may write $z(s) = ct + \int_0^s \int_{\mathbf{R}^d 0} \zeta(N(dud\zeta) - 1_{\{|\zeta|\le 1\}}.\mu(d\zeta))$, where N is a Poisson random measure with mean $du \times \mu(d\zeta)$. Let $\gamma(x,\zeta) : \mathbf{R}^d \times \mathbf{R}^d \to \mathbf{R}^d$ a C^∞-function, whose derivatives of all orders are bounded, satisfying $\gamma(x,0) = 0$.

We carry out our study in the probability space $(\Omega, (\mathcal{F}_t)_{t\ge 0}, P)$, where $\Omega = D[\mathbf{R}^+, \mathbf{R}^d]$ (Skorohod space with the Skorohod topology), $(\mathcal{F}_t)_{t\ge 0}$ = filtration generated by $z(s)$, P = probability measure on ω of $z(s)$. That is, $\mu(d\zeta) = P(z(s+ds) \in d\zeta | z(s))/ds$.

Consider the SDE

$$x_t(x) = x + \int_0^t b(x_s(x))ds + \sum_{s\le t}^{c} \gamma(x_{s-}(x), \Delta z(s)), \tag{1.1}$$

where $\sum^c$ denotes the compensated sum. That is,

$$\sum_{s\le t}^{c} \gamma(x, \Delta z(s)) = \lim_{\epsilon\to 0}\Big\{ \sum_{s\le t, |\Delta z(s)|\ge\epsilon} \gamma(x, \Delta z(s)) - \int_0^t ds \int_{|\zeta|\ge\epsilon} \gamma(x,\zeta)\mu(d\zeta)\Big\}.$$

Equivalently, we may write $x_t(x)$ as

$$x_t(x) = x + \int_0^t b'(x_s(x))ds + \int_0^t \int_{|\zeta|\le 1} \gamma(x_{s-}(x),\zeta)\tilde{N}(dsd\zeta) \tag{1.2}$$

$$+ \int_0^t \int_{|\zeta|>1} \gamma(x_{s-}(x),\zeta)N(dsd\zeta),$$

where $\tilde{N}$ denotes the compensated Poisson random measure : $\tilde{N}(dsd\zeta) = N(dsd\zeta) - ds\mu(d\zeta)$, $b'(x) - b(x) - \int_{|\zeta|\ge 1}\gamma(x,\zeta)\mu(d\zeta)$ where the integrability of $\gamma(x,\zeta)$ with respect to $1_{\{|\zeta|\ge 1\}}.d\mu(\zeta)$ is assumed. We remark that $\gamma(x,\zeta) = \frac{\partial\gamma}{\partial\zeta}(x,0)\zeta + \tilde{\gamma}(x,\zeta)$ for some $\tilde{\gamma}(x,\zeta) = o(|\zeta|)$ as $\zeta\to 0$.

Throughout this paper we adopt the following four assumptions due to Picard [21] ((A.1) $\sim$ (A.4)):

(A.1) the Lévy measure assures that there exists some $0<\beta<2$ and positive C_1, C_2 such that , as $\rho\to 0$,

$$C_1\rho^{2-\beta} I \le \int_{|\zeta|\le\rho} \zeta\zeta^*\mu(d\zeta) \le C_2\rho^{2-\beta} I. \tag{1.3}$$

($\zeta = (\zeta_1,...,\zeta_d)$ is considered as a column vector here and hence inequalities are in the sense of matrices). Or, equivalently, as $\rho\to 0$, for all $u\in S^{d-1}$, $\int_{|\zeta|\le\rho} <\zeta,u>^2 \mu(d\zeta) \asymp \rho^{2-\beta}$. (Here $\asymp$ means the quotient of the two sides is bounded away from zero and above as $\rho\to 0$.)

(A.2) We further adopt the following assumptions in case $0<\beta\le 1$ in the above:

(2-a) If $0<\beta<1$, we assume $c = \int_{|\zeta|\le 1}\zeta\mu(d\zeta)$, $b=0$ and, as $\rho\to 0$, for all $u\in S^{d-1}$,

$$\int_{\{|\zeta|\le\rho\}} <\zeta,u>^2 1_{\{<\zeta,u>>0\}}(\zeta)\mu(d\zeta) \asymp \rho^{2-\beta}. \tag{1.4}$$

(2-b) If $\beta = 1$, then

$$\limsup_{\epsilon\to 0}\Big|\int_{\{\epsilon<|\zeta|\le 1\}} \zeta\mu(d\zeta)\Big| < +\infty. \tag{1.5}$$

(A.3) (3-a) For any $p\ge 2$ and any $k\in\mathcal{N}^d\setminus\{0\}$,

$$\int |\gamma(x,\zeta)|^p d\mu(\zeta) \le C(1+|x|)^p, \quad \sup_x \int \Big|\frac{\partial^k\gamma}{\partial x^k}(x,\zeta)\Big|^p d\mu(\zeta) < +\infty.$$

(3-b) There exists $\Delta > 0$ such that

$$\inf\Big\{y^*\Big(\frac{\partial\gamma}{\partial\zeta}(x,0)\Big)\Big(\frac{\partial\gamma}{\partial\zeta}(x,0)\Big)^* y; x \in \mathbf{R}^d\Big\} \geq \Delta|y|^2 \tag{1.6}$$

on $\mathbf{R}^d$.

(A.4) We assume that, for some $C > 0$,

$$\inf_{x\in\mathbf{R}^d,\zeta\in supp\mu}\Big|\det(I + \frac{\partial\gamma}{\partial x}(x,\zeta))\Big| > C. \tag{1.7}$$

The condition (1.7) guarantees the existence of the flow $\phi_{st}(x)(\omega) : \mathbf{R}^d \to \mathbf{R}^d, x_s(x) \mapsto x_t(x)$ of diffeomorphisms for all $0 < s \leq t$ (cf. [16]).

Equation (1.1) ((1.2)) has a unique solution. This is since (1.1) can be written in the canonical form (in the *Ito integral*)

$$dx_t(x) = d\Theta_t(x_{t-}(x)), \quad x_0(x) = 0, \tag{1.8}$$

where

$$\Theta_t(y) = b'(y)t + \int_0^t \Big\{\int_{|\zeta|\leq 1} \gamma(y,\zeta)\tilde{N}(dsd\zeta) + \int_{|\zeta|>1} \gamma(y,\zeta)N(dsd\zeta)\Big\},$$

and since the equation (1.8) has a unique solution due to [6] Theorem 2.1 (cf. [15] Theorem 3.1). Further, due to the inversibility assumption (A.4), there exists a stochastic flow of diffeomorphism, written $\phi_{s,t}(s < t) : \mathbf{R}^d \to \mathbf{R}^d$, such that $x_t(x) = \phi_{s,t}(x_s(x))$, and the flow is inversible ([16] Section 1, [1] Lemma 7-27; see also [25] Theorem V.65 for the simple case $\gamma(x,\zeta) = X(x)\zeta$).

We state the fundamental result due to Picard [21].

Proposition 1.1. *(cf. [21] Theorem 4.1) Under the conditions (A.1)–(A.4), $x_t(x)$ has a C_b^∞-density for each $t > 0$, which we denote by $y \mapsto p_t(x,y)$.*

We remark the crucial condition:

for each $t > 0, p \geq 1$,

$$\Big\|(\det(\int_0^t \psi_s\psi_s^* ds))^{-1}\Big\|_{L^p} < +\infty \tag{1.9}$$

with $\psi_s = (\frac{\partial}{\partial x}\phi_{s,t}(x_s(x)))(\frac{\partial\gamma}{\partial\zeta}(x_s(x),0))$,
for the existence of the density in [21] is derived from (1.6) and (1.7) by an argument similar to that stated in the proof of corollary 4.4 of [21].

Let $\nu(d\zeta)$ be the probability measure on $\mathbf{R}^d$ given by

$$\nu(d\zeta) = \frac{(|\zeta|^2 \wedge 1)\mu(d\zeta)}{\int(|\zeta|^2 \wedge 1)\mu(d\zeta))}. \tag{1.10}$$

Then $d\nu \sim d\mu$ and $\frac{d\nu}{d\mu}$ is globally bounded from above.

The following result is an extension of Theorem 1 in [23] to the above mentioned type of processes.

Theorem 1.2. *(general upper bound : cf. Picard [23]) The density $p_t(x,y)$ satisfies the following estimate :*

$$(a) \qquad \sup_{x,y} p_t(x,y) \le C_0 t^{-\frac{d}{\beta}} \quad as \ \ t \to 0, \tag{1.11}$$

$$p_t(x,x) \quad \asymp \quad t^{-\frac{d}{\beta}} \quad as \ \ t \to 0 \quad uniformly\ in \quad x. \tag{1.12}$$

(b) *For all $k \in \mathbf{N}^d$ there exists $C_k > 0$ such that*

$$\sup_{x,y} |p_t^{(k)}(x,y)| \le C_k t^{-\frac{(|k|+d)}{\beta}} \quad as \ \ t \to 0 \tag{1.13}$$

where $p^{(k)}$ denotes the k-th derivative with respect to y.

We can give examples that the supremum in (a) is in fact attained on the diagonal $\{x = y\}$ (diagonal estimate) :

$$p_t(x,x) \quad \asymp \quad t^{-\frac{d}{\beta}} \quad \text{as} \ \ t \to 0$$

uniformly in x. See [9], Section 2–4. The estimate above is more exact than the one by Hoh–Jacob [7] using functional analytic methods : there exist $C > 0, \nu \in (1,\infty)$ such that

$$\sup_{x,y} p_t(x,y) \le C t^{-\nu} \quad \text{as} \quad t \to 0.$$

We omit the proof of this theorem, since it is given in [13] Section 3.

Consider a series of functions $(A_n)_{n=0}^{\infty}$, $A_n : \mathbf{R}^{d\times(n+1)} \to \mathbf{R}^d$, defined by

$$\begin{cases} A_0(x) = x, \\ A_{n+1}(x, x_1, ..., x_{n+1}) = A_n(x, x_1, ..., x_n) + \gamma(A_n(x, x_1, ..., x_n), x_{n+1}). \end{cases}$$

Fix $x \in \mathbf{R}^d$. We take $\mathcal{S}_n$ to be the support of the image measure of $\mu^{\otimes n}$ by the mapping $(x_1, ..., x_n) \mapsto A_n(x, x_1, ..., x_n)$, and $\mathcal{S} \equiv \bigcup_n \mathcal{S}_n$.

Definition 1.3. *(accessible points) Points in $\mathcal{S}$, regarded as points in $\mathbf{R}^d$, are called accessible points. Points in $\overline{\mathcal{S}} \setminus \mathcal{S}$ are called asymptotically accessible points.*

Intuitively, accessible points are those points which can be reached by $x_t(x)$ using only a finite number of jumps of $z(s)$. We remark that $\mathcal{S}$ is not necessarily closed, although each $\mathcal{S}_n$ is.

We put for each x the mapping $H_x : \text{supp } \nu \to P_x \equiv x + \{\gamma(x,\zeta);\ \zeta \in \text{supp } \nu\}$, $\zeta \mapsto x + \gamma(x,\zeta)$. Let $P_x^{(n)} = \{y \in P_{z_{n-1}}; z_1 \in P_x, z_i \in P_{z_{i-1}}, i = 2, ..., n-1\}$, $n = 1, 2, ...$ $(z_0 = x)$. Then $P_x^{(1)} = P_x$, and $P_x^{(n)}$ can be interpreted as points which can be reached from x by n jumps along the trajectory $x_t(x)$. Given $x, y \in \mathbf{R}^d (y \neq x)$, let $\alpha(x,y)$ be the minimum number l such that $y \in P_x^{(l)}$ if such l exists, and put $\alpha(x,y) = +\infty$ if not. Or equivalently, $\alpha(x,y) = \inf\{n; y \in \cup_{k\le n} \mathcal{S}_k\}$.

Now we introduce a concrete "singular" Lévy measure of $z(s)$, which already has been described in [27] Example 3.7, and in [2] Section 2. Let $\mu(d\zeta) = \sum_{n=0}^{\infty} k_n \Delta_{a_n}(d\zeta)$ be the d-dimensional Lévy measure such that $(a_n; n \in \mathbf{N})$ and let $(k_n; n \in \mathbf{N})$ be sequences of points in $\mathcal{R}^d$ and real numbers respectively satisfying

- $|a_n|$ decreases to 0 as $n \to \infty$,
- $k_n > 0$,
- $\sum_{n=0}^{\infty} k_n |a_n|^2 < +\infty$.

For this Lévy measure, we can show the unique existence of the solution $x_t(x)$ of (1.2) ([27] Theorems 1.1, 2.1), and the existence of the density under the assumptions (A.1)–(A.4). We further assume

$$N = N(t) \equiv \max\{n; |a_n| > t^{1/\beta}\} \asymp \log(\frac{1}{t}). \tag{1.14}$$

Before stating the following theorem, we give the definition of the functional $\Gamma = \Gamma(t)$ which is used below. Choose any $\beta' > \beta$. Let

$$\Gamma \equiv -\min \sum_{n=0}^{N} (w_n \log(\frac{1}{tk_n}) + \log(w_n!)). \tag{1.15}$$

Here the minimum is taken with respect to all choices of $a_0, ..., a_N$ by x_n for $n = 1, 2, ..., n_1$ and $n_1 \in \mathbf{N}$ such that

$$|y - C_{n_1}(x, x_1, ..., x_{n_1})| \le t^{1/\beta'}.$$

Here $w_n = \#$ of a_n in the choice and $n_1 = \sum_{n=0}^{N} w_n$.

Part (b) of the next theorem can be viewed as an extension of Proposition 5.2 in [22].

Theorem 1.4. *Let $y \neq x$.*

(a) Assume $y \in \mathcal{S}$, that is, $\alpha(x, y) < +\infty$. Then we have

$$p_t(x, y) \quad \asymp \quad t^{\alpha(x,y)-d/\beta} \qquad as \ \ t \to 0. \tag{1.16}$$

(b) Assume $y \in \overline{\mathcal{S}} \setminus \mathcal{S}$ $(\alpha(x, y) = +\infty)$. Suppose $b(x) \equiv 0$ and let $\beta' > \beta$. Then $\log p_t(x, y)$ is bounded from above by an expression of type

$$\Gamma(t) + O(\log(\frac{1}{t}) \log\log(\frac{1}{t})) \tag{1.17}$$

as $t \to 0$.

(c) Assume $y \in \overline{\mathcal{S}} \setminus \mathcal{S}$ $(\alpha(x, y) = +\infty)$. Then $\log p_t(x, y)$ is bounded from below as $t \to 0$ by the expression of type

$$-(c/\beta)(\log(\frac{1}{t}))^2 + O(\log(\frac{1}{t}))$$

with some $c > 0$ independent of β and y.

We remark that in finding the minimum the above conditions (1.14) and (1.15) work complementarily. That is, as $t > 0$ gets smaller, (1.15) becomes apparently more strict, whereas we may use more combinations of a_i's to approximate y due to (1.14). Since $\beta' > \beta$, the condition (1.15) does not prevent the attainability of $A_{n_1}(x, \xi_1, ..., \xi_{n_1})$ to y as $t \to 0$, using $a_0, ..., a_N$ under (1.14). We also remark that if x and a_n are rational points (e.g., $d = 1, x = 0, a_n = 2^{-n}$), then the result (b) holds for almost all $y \in \overline{S}(= [0,1])$ relative to the Lebesgue measure, whereas for $t > 0$, $y \mapsto p_t(x, y)$ is smooth on $\overline{S}$ due to Proposition 1.1. The value of $\Gamma(t)$ for $t > 0$ small (and the value of (1.17) as a whole for $t > 0$ small) depends implicitly on the objective point y.

The proof of this theorem will be given in Sections 3, 4.

Notation Here we sum up notation we will use in the paper.

$z(s)$ Lévy process with Lévy measure $d\mu$

$x_s(x)$ the jump type process driven by $z(s)$

$\tilde{z}(s)$ Lévy process with Lévy measure ν

Z_s Lévy process with Lévy measure $\mu - \nu$

$\tilde{Z}_s^r$ Lévy process defined by $\tilde{Z}_s^r = \sum_{u \le s} \Delta Z_u 1_{\{|\Delta Z_u| > r\}}$

Z_s^r Lévy process defined by $Z_s^r = Z_s - \tilde{Z}_s^r$

$\tilde{z}^r(s)$ Lévy process defined by $\tilde{z}^r(s) = \sum_{u \le s} \Delta z(u) 1_{\{|\Delta z(u)| > r\}}$

$z^r(s)$ Lévy process defined by $z^r(s) = z(s) - \tilde{z}^r(s)$

$x_t^{(r)}(x)$ solution to the equation

$$x_t^{(r)}(x) = x + \int_0^t b(x_{s-}^{(r)}(x))ds + \sum_{s \le t}^{c} \gamma(x_{s-}^{(r)}(x), \Delta(\tilde{z}(s) + Z_s^r))$$

$\tilde{x}_t(x)$ the jump type process driven by

$$\tilde{z}(s) : \tilde{x}_t(x) = x + \sum_{s \le t} \gamma(\tilde{x}_{s-}(x), \Delta\tilde{z}(s))$$

2. Preliminaries

In this section, we shall state several lemmas which are essential for our estimation. These lemmas are inspired by Picard [23]. Proofs of these lemmas can be derived easily in view of Picard [23], using several corresponding lemmas. Let $(\xi_n)_{n \in \mathbf{N}}$ be $\mathbf{R}^d$-valued series of random variables (i.i.d.) obeying the probability law $\nu(d\zeta)$, independent of $z(s)$.

Lemma 2.1. *(lower bound for accessible points) Let $(\xi_n)_{n \in \mathbf{N}}$ be $\mathbf{R}^d$-valued random variables (i.i.d.) obeying the probability law $\nu(d\zeta)$, independent of $z(s)$. We put a Markov chain $(U_n)_{n \in \mathbf{N}}$ by $U_0 = x, U_{n+1} = U_n + \gamma(U_n, \xi_{n+1}), n \in \mathbf{N}$. Assume for $y \in \mathbf{R}^d$, there exist some $n \ge 1, \gamma = \gamma_n \ge 0$ and $c > 0$ such that for all $\epsilon \in (0,1]$ $P(|U_n - y| \le \epsilon) \ge c\epsilon^\gamma$. Then we have*

$$p_t(x, y) \ge Ct^{n + (\gamma - d)/\beta} \quad \textit{as} \quad t \to 0. \tag{2.1}$$

We notice the lower bound in R.H.S. depends on (n, γ_n). Put $g(x, dz) = d(H_x^*\nu)(z), z \in P_x \setminus \{x\}$, where $H_x^*\nu = \nu \circ H_x^{-1}$, $H_x : \zeta \mapsto x + \gamma(x, \zeta)$. Then we have an expression of the probability above :

$$P(|U_n - y| \le \epsilon) = \int_{P_x} \cdots \int_{P_{z_{n-1}}} 1_{\{z_n;|z_n-y|\le\epsilon\}}(z_n) g(x, dz_1)...g(z_{n-1}, dz_n) \quad (2.2)$$

(cf. [10]). Hence the condition $P(|U_n - y| \le \epsilon) \ge c\epsilon^\gamma$ implies :

• *y can be attained with the singular Lévy measure* (dim supp$\nu = 0$) if $\gamma = 0$.

Let $(\varphi_n)_{n\in\mathbf{N}}$ be a series of smooth functions : $\mathbf{R}^d \to \mathbf{R}^d$. We define another Markov chain $(V_n)_{n\in\mathbf{N}}$ by $V_0 = \varphi_0(x), V_{n+1} = V_n + (\varphi_{n+1} \circ \gamma)(V_n, \xi_{n+1})$. Further we set the series of real numbers $(\Phi_n)_{n\in\mathbf{N}}$ by

$$\Phi_n \equiv \sup_{k\le n, y\in\mathbf{R}^d} (|\varphi_k(y) - y| + |\varphi_k'(y) - I|). \quad (2.3)$$

Under these preparations we have

Lemma 2.2. *(upper bound) Choose $y \ne x$. Assume there exist a sequence $(\gamma_n)_{n\in\mathbf{N}}$, $\gamma_n \in [0, \infty]$, and a non-decreasing sequence $(K_n)_{n\in\mathbf{N}}, K_n > 0$ satisfying the following condition: for each n, for any $(\varphi_k)_{k=0}^n$ satisfying $\Phi_n \le K_n$, V_n defined as above satisfy with some $C_n > 0$ the conditions:*

$$\text{If } \gamma_n < +\infty, \quad \text{then} \quad P(|V_n - y| \le \epsilon) \le C_n\epsilon^{\gamma_n} \quad \text{for all} \quad \epsilon > 0. \quad (2.4)$$

$$\text{If } \gamma_n = +\infty, \quad \text{then} \quad P(|V_n - y| \le \epsilon) = 0 \ \text{ for } \quad \epsilon > 0 \quad \text{small.} \quad (2.5)$$

Further we put $\Gamma \equiv \min_n(n + (\gamma_n - d)/\beta)$. Then we have;

If $\Gamma < +\infty$, then $p_t(x, y) = O(t^\Gamma)$ as $t \to 0$.

If $\Gamma = +\infty$, then for any $n \in \mathbf{N}$ $p_t(x, y) = o(t^n)$ as $t \to 0$.

Note that Γ depends implicitly on the choice of (K_n). However, for each n, the bigger γ_n gives the better upper bound.

We take further, given perturbations (φ_j), $Q_x^{(0)} = \{\varphi_0(x)\}$, and the sequence $(Q_x^{(n)})$ successively by

$$Q_x^{(n)} \equiv \{z_{n-1} + (\varphi_n \circ \gamma)(z_{n-1}, \zeta); \zeta \in \text{supp}\nu, z_{n-1} \in Q_x^{(n-1)}\}$$

for $n = 1, 2, ...$ Hence the set $Q_x^{(n)}$ can be interpreted as the points which can be reached from x by V_n, and we have

$$\begin{aligned} P(|V_n - y| \le \epsilon) &= \textstyle\int_{P_{\varphi_0(x)}} \cdots \int_{P_{\varphi_{n-1}(z_{n-1})}} 1_{\{z_n;|\varphi_n(z_n)-y|\le\epsilon\}}(z_n) \\ &\quad g(\varphi_0(x), dz_1)...g(\varphi_{n-1}(z_{n-1}), dz_n) \end{aligned} \quad (2.6)$$

$n = 0, 1, 2, ...$ We remark that if $y \notin P_x^{(n)}$ $(n \ge 1)$, then, by choosing such φ_n that the size Φ_n of perturbations is small enough, we can let $y \notin Q_x^{(n)}$; so that by choosing $K_n > 0$ small and $\epsilon > 0$ small, γ_n may be $+\infty$ in the above.

Some examples of the preceeding results can be found in [13] Section 2.

Let $x_t^{(r)}(x)$ be the process driven by small jumps $\tilde{z}(s) + Z_s^r$ given by

$$x_t^{(r)}(x) = x + \int_0^t b(x_{s-}^{(r)}(x))ds + \sum_{s\le t}^{c} \gamma(x_{s-}^{(r)}(x), \Delta(\tilde{z}(s) + Z_s^r)).$$

We also denote by $\tilde{x}_t(x)$ the process given by

$$\tilde{x}_t(x) = x + \sum_{s\le t}^{c} \gamma(\tilde{x}_{s-}(x), \Delta\tilde{z}(s)).$$

Lemma 2.3. *There exists a matrix-valued function A such that*

$$x_t^{(r)}(x) = \tilde{x}_t(x) + \sum_{s\le t}^{c} A(s, \tilde{z}(.))\frac{\partial}{\partial x}\gamma(\tilde{x}_s(x), 0)\Delta Z_s^r + o(t^{\frac{1}{\beta}}).$$

In particular, $x_t^{(r)}(x) - \tilde{x}_t(x)$ is of order $t^{\frac{1}{\beta}}$.

Lemma 2.4. *Consider an $\mathbf{R}^d$-valued infinitely divisible random variable Θ_t indexed by $t > 0$, and another $\mathbf{R}^d$-valued random variable Y_t. We assume:*

(i) Y_t has a C_b^∞ transition density uniformly in t.

(ii) $\|Y_t - \Theta_t\|_{L^1} \to 0$ as $t \to 0$.

(iii) We denote by μ_t the Lévy measure of Θ_t. There exist a compact $K \subset \mathbf{R}^d$, and $M > 0$ such that $\text{supp}\mu_t \subset K$, and $|x|^2\mu_t(dx) \le M$ (uniformly in $t \le 1$), respectively.

Further we assume μ_t satisfies (A.1) and (A.2). Then the density of Y_t is bounded away from zero as $t \to 0$ on any compact set.

For proof of the last two lemmas, see [23] Lemma 8, Lemma 6. The conclusion of Lemma 2.4 follows essentially from the uniform boundedness of $\mu_t(dx)$.

3. Proof of Theorem 1.4 (a), (b)

For completeness in comparing results according to the cases in Theorem 1.4, we repeat in this section some arguments in [13].

Proof of (a). For $n \ge \alpha(x, y)$ we have $P(|U_n - y| \le \epsilon) \ge c$ for $\epsilon \in (0, 1]$, since μ has point masses. Hence $p_t(x, y) \ge Ct^{\alpha(x,y)-d/\beta}$ by Lemma 2.1.

For the upper bound, if $n < \alpha(x, y)$, then by choosing small K_n we have $y \notin Q_x^{(n)}$. That is,

$$P(|V_n - y| \le \epsilon) = 0 \quad \text{for } \epsilon > 0 \text{ small},$$

and we may choose $\gamma_n = +\infty$ in (2.4), (2.5). On the other hand, if $n \ge \alpha(x, y)$, then we must choose $\gamma_n = 0$ in (2.4) ($\varphi_n = id$ must satisfy it).

Hence we may choose $\Gamma = \alpha(x, y) - d/\beta$ in the conclusion. These imply

$$p_t(x, y) \asymp t^{\alpha(x,y)-d/\beta}. \tag{3.1}$$

Proof of (b). We set $\mathbf{S}_{t,k} = ([0,t]^k/\sim)$ and $\mathbf{S}_t = \amalg_{k\geq 0}\S_{t,k}$, where $\amalg_{k\geq 0}$ denotes the disjoint sum and $\sim$ means the identification of the coordinates on the product space $[0,t]^k$ by the permutation.

Let $r = r(t) = t^{1/\beta}$. The distribution $\widetilde{P}_{t,r}$ of the moments (instants) of jumps related to $\tilde{z}^r(s)$ during $[0,t]$ is given by

$$\begin{aligned}&\int_{\{\#S_t=k\}} f(S_t)\, d\tilde{P}_{t,r}(S_t)\\ &=\left\{\left(t\int 1_{\{|\zeta|>r\}}(\zeta)\mu(d\zeta)\right)^k\left(\frac{1}{k!}\right)\exp\left(-t\int 1_{\{|\zeta|>r\}}(\zeta)\mu(d\zeta)\right)\right\}\\ &\quad\times\exp\left(-t\int 1_{\{|\zeta|>r\}}(\zeta)\mu(d\zeta)\right),\end{aligned}\tag{3.2}$$

where f is a function on $\mathbf{S}_{t,k}$ (a symmetric function on $[0,t]^k$). Given $S_t \in \mathbf{S}_t$, we introduce the process $x_s(r,S_t,x)$ as the solution of the following S.D.E. :

$$\begin{aligned}x_s(r,S_t,x) &= x-\int_0^s du\int 1_{\{|\zeta|>r\}}(\zeta)\gamma(x_u(r,S_t,x),\zeta)\mu^r(d\zeta)\\ &\quad+\sum_{u\leq s}^{c}\gamma\Big(x_{u-}(r,S_t,x),\Delta z^r(u)\Big)\\ &\quad+\sum_{s_i\in S_t, s_i\leq s}\gamma(x_{s_i-}(r,S_t,x),\xi_i^r).\end{aligned}\tag{3.3}$$

Here $(\xi_n^r)_{n\in\mathbf{N}}$ denotes a series of random variables (i.i.d.) obeying the probability law

$$\mu^r(d\zeta) = \frac{1_{\{|\zeta|>r\}}(\zeta).\mu(d\zeta)}{\int 1_{\{|\zeta|>r\}}(\zeta).\mu(d\zeta))}.$$

We remark $x_s(r,S_t,x)$ is a martingale for each $0<r<1$, due to the assumption $b(x)\equiv 0$. We define a new Markov chain $(U_n^r)_{n\in\mathbf{N}}$ by $U_0^r = x, U_{n+1}^r = U_n^r + \gamma(U_n^r,\xi_{n+1}^r), n\in\mathbf{N}$.

We can prove due to Proposition 1.1 that under (A.1)–(A.4) the law of $x_s(r,S_t,x)$ for each $s>0$ has a C_b^∞- density, denoted by $p_s(r,S_t,x,y)$.

Indeed, let $0\leq s\leq t$. In case $S_t=\emptyset$ we can choose $1_{\{|\zeta|\leq r\}}.\mu(d\zeta)$ for the measure $\mu(d\zeta)$ in Proposition 1.1. Hence we have the existence of the density for the law $p_s(r,\emptyset,x,dy)$ of $x_s(r,\emptyset,x)$ $(= x_s(r,S_t,x)|_{S_t=\emptyset})$: $p_s(r,\emptyset,x,dy) = p_s(r,\emptyset,x,y)dy$.

Next we consider the general case. Since $\tilde{z}^r(s)$ and $z^r(s)$ are independent, we have by Markov property that the law $p_s(r,S_t,x,dy)\equiv P(x_s(r,S_t,x)\in dy)$ of $x_s(r,S_t,x)$ is represented by

$$
\begin{aligned}
p_s(r, S_t, x, dy) &= \int dz_0' \int_{P_{z_0'}} \cdots \int dz_{n_1-1}' \int_{P_{z'_{n_1-1}}} \\
&\quad \Big\{ p_{s_1}(r, \emptyset, x, z_0')\, g_r(z_0', dz_1)\, p_{s_2-s_1}(r, \emptyset, z_1, z_1')\, g_r(z_1', dz_2) \\
&\quad \times p_{s_3-s_2}(r, \emptyset, z_2, z_2') \cdots g_r(z_{n_1-1}', dz_{n_1}) \\
&\quad \times p_{t-s_{n_1}}(r, \emptyset, z_{n_1}, y) \Big\} \qquad (3.4)
\end{aligned}
$$

if $S_t \in \mathbf{S}_{t,n_1}$ (cf. [10] (2.7)). Here $g_r(x, dz) = P(x + \gamma(x, \xi_i^r) \in dz)$ and $P_z = z + \{\gamma(x, \zeta); \zeta \in \text{supp}\mu^r\}$.

On the other hand, we have by the independence of $\tilde{z}^r(s)$ and $z^r(s)$ again

$$
p_s(x, dy) = \int_{\mathbf{S}_t} p_s(r, S_t, x, dy) d\tilde{P}_{t,r}(S_t)
$$

using the factorization of the measure λ^+ (cf. [14] p. 71). By Proposition 1.1, L.H.S. has the density $p_s(x, y)$ with respect to dy. Hence $p_s(r, S_t, x, dy)$ is absolutely continuous with respect to dy ($d\tilde{P}_{t,r}$-a.s.). Hence we have by the derivation under the integral sign

$$
p_s(x, y) = \int_{\mathbf{S}_t} (p_s(r, S_t, x, dy)/dy)(y) d\tilde{P}_{t,r}(S_t).
$$

We denote by $p_s(r, S_t, x, y)$ the derivative $p_s(r, S_t, x, dy)/dy(y)$ which is defined uniquely $d\tilde{P}_{t,r} \otimes dy$–a.e.. (Since $d\tilde{P}_{t,r}|_{\mathbf{S}_{t,k}}$ is the uniform distribution on $\mathbf{S}_{t,k}$, $p_s(r, S_t, x, y)$ is defined uniquely $ds \otimes dy$–a.e..) Since $y \mapsto p_s(x, y)$ is smooth, so is $y \mapsto p_s(r, S_t, x, y)$ ds-a.s., and $p_s(r, S_t, x, y)$ is defined ds-a.e.. Thus by taking $s = t$,

$$
p_t(x, y) = \int_{\mathbf{S}_t} p_t(r, S_t, x, y) \tilde{P}_{t,r}(S_t) = \sum_{k=0}^{\infty} \tilde{p}_t(k, r, x, y),
$$

where

$$
\tilde{p}_t(k, r, x, y) = \int_{\mathbf{S}_{t,k}} p_t(r, S_t, x, y) d\tilde{P}_{t,r}(S_t).
$$

Hence

$$
\begin{aligned}
p_t(x, y)dy &= E^{\tilde{P}_{t,r}}[P(x_t(r, S_t, x) \in dy)] \\
&= E^{\tilde{P}_{t,r}} E^{(\mu^r)^{\otimes \# S_t}}[P(x_t(r, S_t, x) \in dy | S_t, \xi_1^r, ..., \xi_{\# S_t}^r)]. \qquad (3.5)
\end{aligned}
$$

For each $S_t \in \mathbf{S}_t$, we put $n_1 = k$ if $S_t \in \mathbf{S}_{t,k}$. We have then

$$
\begin{aligned}
&E^{(\mu^r)^{\otimes \# S_t}}[P(x_t(r, S_t, x) \in dy) | S_t] \\
&= E^{(\mu^r)^{\otimes \# S_t}}[P(x_t(r, S_t, x) \in dy : |y - U_{n_1}^r| \le t^{1/\beta'} | S_t, \xi_1^r, ..., \xi_{n_1}^r)] \\
&+ E^{(\mu^r)^{\otimes \# S_t}}[P(x_t(r, S_t, x) \in dy : |y - U_{n_1}^r| > t^{1/\beta'} | S_t, \xi_1^r, ..., \xi_{n_1}^r)]. \qquad (3.6)
\end{aligned}
$$

First we have to calculate, given $y = x_t(r, S_t, x)$ with $S_t \in \mathbf{S}_{t,n_1}$, $\mathbf{P}(|y - U_{n_1}^r| \le t^{1/\beta'}) \equiv E^{(\mu^r)^{\otimes n_1}}[P(|y - U_{n_1}^r| \le t^{1/\beta'} | \xi_1^r, ..., \xi_{n_1}^r)]$. We denote by the random variable W_n the number of a_n in $(\xi_i^r)_{i=1}^{n_1}$.

Given $n_1 \in \mathbf{N}$, let $(w_n)_{n=0}^N, w_n \in \mathbf{N} \cup \{0\}$ be a series of integers such that $n_1 = \sum_{n=0}^N w_n$. We have

$$\mathbf{P}(for\ all\ n \leq N\ (W_n = w_n)\ and\ |y - U_{n_1}^r| \leq t^{1/\beta'}) \\ \leq \Pi_{n=0}^N \frac{1}{w_n!}(tk_n)^{w_n} e^{-tk_n},$$

since each W_n is a Poisson random variable with mean tk_n. Hence

$$\log \mathbf{P}(for\quad all\quad n\quad \leq\quad N\ (W_n = w_n)\quad and\quad |y - U_{n_1}^r| \leq t^{1/\beta'}) \\ \leq\quad -\sum_{n=0}^{N} \big(w_n \log(\frac{1}{tk_n}) + \log(w_n!) + tk_n\big). \tag{3.7}$$

We introduce the set $\mathcal{W}$ of all $(w_n)_{n=0}^N$'s such that for some $n_1 \in \mathbf{N}$, $U_{n_1}^r$ directed by the following condition $(*)$ satisfies $|y - U_{n_1}^r| \leq t^{1/\beta'}$:

$(*)$ $w_n = (\#\ of\ a_n\text{'s which appear in } \xi_1^r, ..., \xi_{n_1}^r)$ *and* $n_1 = \sum_{n=1}^N w_n$.

Then by (3.7)

$$\log \mathbf{P}(there\ exists\ (w_.) \in \mathcal{W}\ such\ that\ for\ all\ n \leq N\ (W_n = w_n)\ and \\ |y - U_{n_1}^r| \leq t^{\frac{1}{\beta'}}) \leq -\min_{\mathcal{W}} \sum_{n=0}^{N} \big(w_n \log(\frac{1}{tk_n}) + \log(w_n!)\big) + O(\log|\mathcal{W}|). \tag{3.8}$$

On the other hand

$$\mathbf{P}(W_n \geq N^3\ \ and\ \ |y - U_{n_1}^r| \leq t^{\frac{1}{\beta'}}) \leq Ce^{-N^3} \leq Ce^{-c(\log(1/t))^3} \tag{3.9}$$

since W_n is a Poisson variable with mean tk_n. Since this is very small relative to the probability above, we may put the restriction $w_n \leq N^3$ and $n_1 \leq N^4$. Hence we may write $|\mathcal{W}| = O((N^3)^N)$,

$$\log|\mathcal{W}| = O(N \log N) = O(\log(1/t) \log\log(1/t)).$$

Let $x_t(r, \emptyset, x)$ denote the process defined by (3.3) with $S_t = \emptyset$, $p_t(r, \emptyset, x, y)$ its density. By Theorem 1.2 (a) $p_t(r, \emptyset, x, y) \leq Ct^{-d/\beta}$ as $t \to 0$, and this implies, given $\xi_1^r, ..., \xi_{n_1}^r$, that

$$E^{\tilde{P}_{t,r}|\mathbf{s}_{t,n_1}}[P(x_t(r, S_t, x) \in dy|S_t, \xi_1^r, ..., \xi_{n_1}^r)/dy] = O(t^{-d/\beta})$$

as $t \to 0$. This is since we have, as in (3.4), for $S_t = \{s_1, ..., s_{n_1}\}(s_1 < s_2 < \cdots < s_{n_1})$,

$$P(x_t(r, S_t, x) \in dy|S_t, \xi_1^r, ..., \xi_{n_1}^r)/dy = \int dz_0' \int_{P_{z_0'}} \cdots \int dz_{n_1-1}' \int_{P_{z'_{n_1-1}}} \\ \Big\{p_{s_1}(r, \emptyset, x, z_0')\ \tilde{g}_{r,0}(z_0', dz_1)\ p_{s_2-s_1}(r, \emptyset, z_1, z_1')\ \tilde{g}_{r,1}(z_1', dz_2) \\ \times p_{s_3-s_2}(r, \emptyset, z_2, z_2') \cdots \tilde{g}_{r,n_1-1}(z_{n_1-1}', dz_{n_1})\ p_{t-s_{n_1}}(r, \emptyset, z_{n_1}, y)\Big\}.$$

Here $\tilde{g}_{r,i}(x, dz) = P(x + \gamma(x, J(r)) \in dz | J(r) = \xi_i^r)$, where $J(r)$ is a random variable whose law is μ^r.

Since $z^r(u)$ and (ξ_i^r) are independent, by (3.4) and Fubini's Theorem we have

$$E^{\tilde{P}_{t,r}|\mathbf{S}_{t,n_1}} E^{(\mu^r)^{\otimes n_1}}[P(x_t(r, S_t, x) \in dy : |y - U_{n_1}^r| \le t^{1/\beta'} | S_t, \xi_1^r, ..., \xi_{n_1}^r)]/dy$$
$$\le E^{(\mu^r)^{\otimes n_1}}[Ct^{-d/\beta} P(|y - U_{n_1}^r| \le t^{1/\beta'} | \xi_1^r, ..., \xi_{n_1}^r)]$$
$$\le C't^{-d/\beta} \exp(-\min_{\mathcal{W}} \sum_{n=0}^{N} (w_n \log(\frac{1}{tk_n}) + \log(w_n!))$$
$$\times O(\log(1/t) \log\log(1/t))).$$

Hence

$$\log(E^{\tilde{P}_{t,r}|\mathbf{S}_{t,n_1}} E^{(\mu^r)^{\otimes n_1}}[P(x_t(r, S_t, x) \in dy :$$
$$|y - U_{n_1}^r| \le t^{1/\beta'} | S_t, \xi_1^r, ..., \xi_{n_1}^r)]/dy)$$
$$\le -\min_{\mathcal{W}} \sum_{n=0}^{N} (w_n \log(\frac{1}{tk_n}) + \log(w_n!)) + O(\log(1/t) \log\log(1/t)). \quad (3.10)$$

For the second term, we have the following lemma, whose proof is given in [13] Appendix.

Lemma 3.1. *There exist $k > 0$ and $C_0 > 0$ such that for every $p > k$, for all $y = x_t(r, S_t, x) = x_t(r, S_t, x; \xi_1^r, ..., \xi_{n_1}^r)$ and $U_{n_1}^r = U_{n_1}^r(\xi_1^r, ..., \xi_{n_1}^r) = C_{n_1}(x, \xi_1^r, ..., \xi_{n_1}^r)$,*

$$P(|y - U_{n_1}^r| > t^{1/\beta'} \quad | \quad S_t, \xi_1^r, ..., \xi_{n_1}^r) \le n_1 C_0 \exp[-(p - k)(\log(1/t))^2]$$

as $t \to 0$. Here C_0, k are uniform with respect to $S_t \in \mathbf{S}_{t,n_1}$, $\xi_1^r, ..., \xi_{n_1}^r$ and $0 < r < 1, x \in \mathbf{R}^d$.

We integrate this with respect to $(\mu^r)^{\otimes n_1}$ on $\{|y - U_{n_1}^r| > t^{1/\beta'}\}$ and $\tilde{P}_{t,r}$ on $\mathbf{S}_{t,n_1}$. Since $z^r(u)$ and (ξ_i^r) are independent, we have by Lemma 3.1,

$$E^{\tilde{P}_{t,r}|\mathbf{S}_{t,n_1}} E^{(\mu^r)^{\otimes n_1}}[P(x_t(r, S_t, x) \in dy :$$
$$|y - U_{n_1}^r| > t^{1/\beta'} | S_t, \xi_1^r, ..., \xi_{n_1}^r)]/dy$$
$$\le n_1 C_0' t^{-d/\beta} \exp[-(p - k)(\log(1/t))^2],$$

as $t \to 0$. Then we get

$$\log(E^{\tilde{P}_{t,r}|\mathbf{S}_{t,n_1}} E^{(\mu^r)^{\otimes n_1}}[P(x_t(r, S_t, x) \in dy :$$
$$|y - U_{n_1}^r| > t^{1/\beta'} | S_t, \xi_1^r, ..., \xi_{n_1}^r)]/dy)$$
$$\le -(p - k)(\log(1/t))^2 + \max_{n_1 \le N^4} \log n_1 + \log C_0' + O(\log(\frac{1}{t}))$$
$$\le -(p - k)(\log(1/t))^2 + O(\log(\frac{1}{t})),$$

since $\max_{n_1 \le N^4} \log n_1 = \log N^4 = O(\log(\frac{1}{t}))$. Since $p > k$ is arbitrary, this can be neglected in view of R.H.S. of (3.10) and (3.6).

After summing up $E^{\tilde{P}_{t,r}|\mathbf{s}_{t,n_1}} E^{(\mu^r)^{\otimes n_1}}[...]$ with respect to $n_1 = 0, ..., N^4$, we get

$$\log(E^{\tilde{P}_{t,r}|\cup_{k \le N^4} \mathbf{s}_{t,k}} E^{(\mu^r)^{\otimes \# S_t}}[P(x_t(r, S_t, x) \in dy | S_t, \xi_1^r, ..., \xi_{\# S_t}^r)]/dy)$$
$$\le -\min \sum_{n=0}^{N} (w_n \log(\frac{1}{tk_n}) + \log(w_n!))$$
$$+ O(\log(1/t) \log\log(1/t)) + O(\log(1/t)). \tag{3.11}$$

In view of (3.5), (3.9),

$$\log p_t(x, y) \le -\min \sum_{n=0}^{N} (w_n \log(\frac{1}{tk_n}) + \log(w_n!)) + O(\log(1/t) \log\log(1/t)). \tag{3.12}$$

Since there is no difference between the trajectories of the deterministic chain $A_{n_1}(x, \xi_1, ..., \xi_{n_1})$ and $U_{n_1}^r$ obtained by using $\{a_n; n = 0, ..., N\}$ under (1.14), we have proven the assertion.

4. Proof of Theorem 1.4 (c)

Recall that we denote by $x_s^{(r)}(x)$ the solution of

$$x_s^{(r)}(x) = x + \int_0^s b(x_u^{(r)}(x))du + \sum_{u \le s}^{c} \gamma(x_{u-}^{(r)}(x), \Delta(\tilde{z}(u) + Z_u^r)).$$

The probability that $\tilde{Z}^r \equiv 0$ on $[0, t]$ is bounded below by a positive constant, and $x_t(x) = x_t^{(r)}(x)$ on this set. Hence, in giving the lower bound of $p_t(x, y)$, it is sufficient to estimate the density of $x_t^{(r)}(x)$ at y.

We put $N_1 = \#\{u \le t; |\Delta \tilde{z}(u)| > 0\}$. Let $T_1 < T_2 < \cdots < T_{N_1}$ be its jump times ($T_0 \equiv 0, T_{N_1+1} \equiv t$). We put $A_{n_1} = \{N_1 = n_1\}$. Then

$$P(A_{n_1}) \asymp ct^{n_1}$$

as $t \to 0$.

On the other hand, on A_{n_1},

$$P(x_t^{(r)}(x) \in dy | \tilde{z}(.)) = P(\tilde{x}_t^{(r)}(x) \in dy | \tilde{z}(.)) \tag{4.1}$$

where $\tilde{x}_s^{(r)}(x)$ is given by

$$\tilde{x}_s^{(r)}(x) = x + \int_0^s b(\tilde{x}_u^{(r)}(x))du + \sum_{u \le s}^{c} \gamma(\tilde{x}_{u-}^{(r)}(x), \Delta Z_u^r). \tag{4.2}$$

Then by Proposition 1.1, $x_t^{(r)}(x)$ has a C_b^∞-density conditioned by $\tilde{z}(.)$. Indeed, the conditions (1.4), (3.3) of [21] are satisfied by assumptions (A.1), (A.4) respectively. We apply the process $t' \mapsto \bar{x}^{(r)}_{t'}(x)$ by $\bar{x}^{(r)}_{t'}(x) \equiv \frac{1}{r}x^{(r)}_{r^\beta t'}(x)$. In view of $r = r(t) = t^{\frac{1}{\beta}}$, we deduce that the infinitely divisible random variable

$$\bar{x}^{(r)}_1(x) \equiv \tilde{x}^{(r)}_t(x)/t^{\frac{1}{\beta}}$$

given $\tilde{z}(.)$ has a C_b^∞-density on A_{n_1}. (Proposition 1.1 applied at $t' = 1$.) Since $\tilde{x}_s(x)$ is driven by $\tilde{z}(.)$, $\frac{1}{t^{1/\beta}}(x^{(r)}_t(x) - \tilde{x}_t(x))$ satisfies the same property. However, by Lemma 2.3, it is equivalent (conditionally on $\tilde{z}(.)$) to some infinitely divisible random variable.

By Lemma 2.4, there exists $c > 0$ such that, on A_{n_1},

$$P\Big(\frac{1}{t^{1/\beta}}(x^{(r)}_t(x) - \tilde{x}_t(x)) \in dz|\tilde{z}(.)\Big) \geq cdz. \tag{4.3}$$

Hence, on $\{\omega : |y - \tilde{x}_t(x)| \leq t^{\frac{1}{\beta}}\} \cap A_{n_1}$,

$$P(x^{(r)}_t(x) \in dy|\tilde{z}(.)) \geq c' dy t^{-\frac{d}{\beta}}. \tag{4.4}$$

Hence

$$P(x^{(r)}_t(x) \in dy) \geq c' t^{-\frac{d}{\beta}} P(|\tilde{x}_t(x) - y| \leq t^{\frac{1}{\beta}} : A_{n_1}) dy. \tag{4.5}$$

On the other hand, the jumps $(\Delta\tilde{z}_{T_i})_{i=1,\dots,n_1}$ are mutually independent, having the same law ν, and are independent of T_i. The random variables (ξ_n) are i.i.d. variables, following the same law ν, and are independent of T_i.

Hence $\tilde{x}_t(x)$ and U_{n_1} have the same law on A_{n_1}, and

$$P(|\tilde{x}_t(x) - y| \leq t^{1/\beta} : A_{n_1}) = P(|U_{n_1} - y| \leq t^{1/\beta} : A_{n_1}). \tag{4.6}$$

Next we shall compute the probability $P(|U_{n_1} - y| \leq t^{1/\beta})$ for n_1 large. We recall $r = r(t) = t^{1/\beta}$. Given $y \in \bar{\mathcal{S}} \setminus \mathcal{S}$, there exists a series $(y_j)_{j\in\mathbf{N}}, y_j \in \mathcal{S}$, such that $y_0 = x$ and $y_j \to y$ as $j \to \infty$. We may assume $y_j \in \cup_{k\leq j}\mathcal{S}_k$. We denote by $\mathcal{V}$ the set of such series.

Given $(y_j) \in \mathcal{V}$, let $J = J(\epsilon) = \min\{j; |y - y_j| < \epsilon\}$. Note that $J \to \infty$ as $\epsilon \to 0$.

We will choose ϵ in the form $\epsilon = \epsilon(t)$ so that $\epsilon(t) \to 0$ and $\epsilon(t) \asymp t^{1/\beta}$ as $t \to 0$. We may assume, given $(y_j) \in \mathcal{V}$, that we can choose an integer n_1 so that

$$|y_{n_1} - y| < \epsilon \quad \text{and} \quad n_1 \asymp N(t)$$

by the definition of $N(t)$ in (1.14). Obviously, $n_1 \geq J(\epsilon)$.

Since $(y_j) \in \mathcal{V}$, we have $y_{n_1} \in \mathcal{S}$. We can show that there exists a step function $\phi_{n_1}(s)$ such that

$$\phi_{n_1}(s) = x + \int_0^s b(\phi_{n_1}(s'))ds' + \sum_{t_i\leq s}\gamma(\phi_{n_1}(t_i-), z_i), \quad \phi_{n_1}(t) = y_{n_1}.$$

Indeed, define $\phi_{n_1}(s)$ by $\phi_{n_1}(s) = y_k$, if $s \in [(k/n_1)t, ((k+1)/n_1)t)$, and $\phi_{n_1}(s) = y_{n_1}$, if $s = t$. Then $\phi_{n_1}(s)$ has the above form with $t_k = (k/n_1)t$, and $z_k = (H_{y_{k-1}} - y_{k-1})^{-1}(y_k - y_{k-1})$.

We put $u_i \equiv (t_i, z_i) \in [0, t] \times \mathbf{R}^m, i = 1, \dots, n_1$, and put

$$\eta \equiv (1/2)\inf\{|z_i|; i = 1, \dots, n_1\}.$$

Further we write below ϕ for ϕ_{n_1} for simplicity. We denote by $x^\eta_\cdot(x)$ trajectories given by

$$x^\eta_s(x) = x + \int_0^s b(x^\eta_{s'}(x))ds' + \sum_{\tilde{T}_i \le s} \gamma(x^\eta_{\tilde{T}_i -}(x), Z_i), \quad x^\eta_0(x) = x.$$

Here $\tilde{T}_1, \tilde{T}_2, \ldots$ are jump times of $\tilde{z}(.)$ such that

$$|\Delta \tilde{z}(.)| > \eta \text{ and } Z_i = \Delta \tilde{z}(\tilde{T}_i).$$

Then

Lemma 4.1. *There exist $\rho > 0, \delta > 0$, and $\Omega_1 = \Omega_1(\rho, \delta) \subset A_{n_1}$ such that*

$$P(\Omega_1) > 0 \quad and \quad \mathbf{d}_t(\phi, x^\eta_\cdot) < \epsilon \tag{4.7}$$

on Ω_1. Here $\mathbf{d}_t$ denotes the Skorohod metric on $[0, t]$.

We construct the time change λ^ρ_t of $[0, t]$ as in [26]. That is, (t^1_i, t^2_i) are given by the following figure [1]

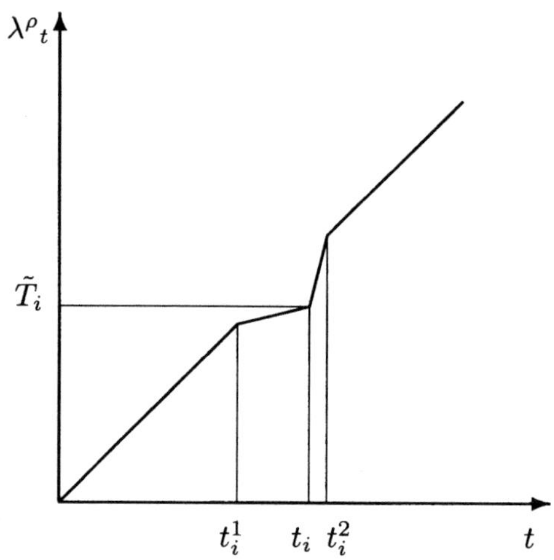

and λ^ρ_t a withe change of $[0, t]$ piecewise affine such that

$$\begin{cases} \lambda^\rho_{t_i} = \widetilde{T}_i \quad i = 1, \ldots, n, \\ \begin{cases} \lambda^{\rho'}_{t_i +} = \rho \\ \lambda^{\rho'}_{t_i -} = 1/\rho \end{cases} \quad if \quad \widetilde{T}_i > t_i, \\ \begin{cases} \lambda^{\rho'}_{t_i +} = 1/\rho \\ \lambda^{\rho'}_{t_i -} = \rho \end{cases} \quad if \quad \widetilde{T}_i < t_i, \\ \lambda^\rho_t = t \qquad on\ every\ \left[t^2_i, t^1_{i+1}\right], \end{cases}$$

where $\lambda^{\rho'}{}_t$ denotes the derivative (from left and right) of λ^ρ.

---

[1] Courtesy of Th. Simon.

Let $\mathbf{U}_i = (\tilde{T}_i, Z_i) \in [0,t] \times {}^m, i = 1, ..., n_1$ and put Ω_1 using $(\mathbf{U}_i)$ by

$$\Omega_1 = \bigcap_{i=1}^{n_1} \left(\left\{ |\tilde{T}_i - t_i| < \rho \right\} \bigcap \left\{ |Z_i - z_i| < \delta \right\} \right) \bigcap \left\{ \tilde{T}_{n_1+1} > t \right\}. \tag{4.8}$$

Then

$$P(\Omega_1) > 0. \tag{4.9}$$

Indeed, Ω_1 contains the set

$$\Omega_2 = \bigcap_{i=1}^{n_1} \left(\left\{ |(\tilde{T}_i - \tilde{T}_{i-1}) - (t_i - t_{i-1})| < \rho/n_1 \right\} \bigcap \left\{ |Z_i - z_i| < \delta \right\} \right)$$
$$\bigcap \left\{ \tilde{T}_{n_1+1} - \tilde{T}_{n_1} > t + \rho - t_{n_1} \right\}.$$

Since $\tilde{T}_i - \tilde{T}_{i-1}$ follows an exponential law for $i = 1, ..., n_1$, and since $z_i \in \mathrm{supp}(\mu|_{\{|\zeta|>2\eta\}})$, Ω_2 has a positive probability as an intersection of independent events.

To show the second assertion, we choose $\rho > 0$ (hence $t > 0$) so that

$$\sup_{0 \le s \le t} |\lambda_s^\rho - s| < \epsilon.$$

This is possible by choosing $\rho = ct^{1/\beta}, 0 < c < (1/\sqrt{2})$, and since

$$\sup_{0 \le s \le t} |\lambda_s^\rho - s| \le \sqrt{2}\rho.$$

On the other hand, note

$$|t_i - t_i^1| \lambda^{\rho'}{}_{t_i -} \;=\; O(\rho), \quad |t_i^2 - t_i| \lambda^{\rho'}{}_{t_i +} \;=\; O(\rho)$$

as $\rho \to 0$, and $t_i^2 - t_i^1 = \sqrt{2}\rho . \frac{1+\rho}{1-\rho} \sim \sqrt{2}\rho$ as $\rho \to 0$.

Let $\psi(s) = x^\eta_{\lambda_s^\rho}(x)$. Then we can show

Lemma 4.2

$$\sup_{0 \le s \le t} |\psi(s) - \phi(s)| < \epsilon.$$

The proof of this lemma is also given in the appendix.

Hence

$$\mathbf{d}_t(\phi, x^\eta_\cdot(x)) < \epsilon.$$

This completes the proof of Lemma 4.1. □

Next we write $\tilde{Y} = \psi(t)$. Then $|\tilde{Y} - y_{n_1}| < \epsilon$ on Ω_1 by the above, hence

$$|y - \tilde{Y}| < \epsilon + \epsilon = 2\epsilon \tag{4.10}$$

on Ω_1.

Note that we have the chain $(\tilde{U}_i)$ by $\psi(s)$ as

$$\begin{cases} \tilde{U}_0 = x, \\ \tilde{U}_i = \psi(t_i) \quad (= x^{\eta}_{\tilde{T}_i}), \quad i = 1, ..., n_1. \end{cases}$$

Then the chain $(\tilde{U}_i)$, corresponding to the chain $(\mathbf{U}_i)$ above through $x^{\eta}_s(x)$, satisfy $\tilde{U}_{n_1} = \psi(t)$ and

$$|y - \tilde{U}_{n_1}| < 2\epsilon \tag{4.11}$$

on Ω_1 by (4.10). Then by the definition of Ω_1,

$$P(|y - \tilde{U}_{n_1}| < 2\epsilon : A_{n_1}) \geq P(\Omega_1(\rho, \delta)). \tag{4.12}$$

Next we calculate $P(\Omega_1)$. We choose δ so that $0 < \delta < \eta$. We put $B(z_i) = \{z; |z - z_i| < \delta\}$. Since $Z_i, i = 1, ..., n_1$ are Poisson random variables with mean $\lambda = t.\nu|_{\{|z|>\eta\}}$, and $\tilde{T}_i$ are their jump times,

$$P(\Omega_1(\rho, \delta)) \geq c'.\Big(\frac{t_1^2 - t_1^1}{t}.\frac{t_2^2 - t_2^1}{t}...\frac{t_{n_1}^2 - t_{n_1}^1}{t}\Big)$$

$$\times\Big\{(t\tilde{\nu}(B(z_1))e^{-t\tilde{\nu}(B(z_1))}) \cdot t\tilde{\nu}(B(z_2))e^{-t\tilde{\nu}(B(z_2))}) \cdots t\tilde{\nu}(B(z_{n_1}))e^{-t\tilde{\nu}(B(z_{n_1}))})\Big\}.$$

Here we put $\tilde{\nu}(A) = \nu|_{\{|z|>\eta\}}(A)$.

Hence

$$P(\Omega_1(\rho, \delta)) \geq (c\rho)^{n_1}, \tag{4.13}$$

for $t > 0$ small, and

$$\begin{aligned} \log RHS &\geq n_1 \log(c\rho) = n_1(\log c + \log \rho) \\ &\asymp n_1 \log c + n_1(1/\beta) \log t \\ &\asymp N(t) \log c + (1/\beta) N(t) \log t \\ &\geq c'(\log(\frac{1}{t}) \log c - (c'/\beta)((\log(\frac{1}{t}))^2), \end{aligned} \tag{4.14}$$

as $t \to 0$, since $n_1 \asymp N(t)$. Here the positive constant $c' > 0$ is controlled by the condition (1.14), and is independent of β and y.

Finally we let $t \to 0$, and $\epsilon \to 0$ accordingly. We have the assertion (c) by (4.14), (4.12), (4.6), (4.5) and the argument at the beginning of this part. □

5. Appendix. Proof of Lemma 4.2

We write

$$\psi(s) = x + \int_0^s b(\psi(s'))\rho_{s'} ds' + \sum_{t_i \leq s} \gamma(x^{\eta}_{\lambda^{\rho}_{t_i-}}(x), Z_i).$$

Then

$$\begin{aligned}|\psi(s)-\phi(s)| &\le K\int_0^s |\psi(s')-\phi(s')|ds' \\ &\quad + \int_0^s |b(\psi(s'))||\rho_{s'}-1|ds' + K\sum_{t_i\le s}(|\psi(t_i-)-\phi(t_i-)|+\delta) \\ &\le Kn_1(\delta+\rho)+K\int_0^s|\psi(s')-\phi(s')|ds' \\ &\quad + K\sum_{t_i\le s}(|\psi(t_i-)-\phi(t_i-)|).\end{aligned}$$

We can choose $\delta=\delta(t)>0$ and ρ small relative to ϵ so that for every $t'\in[0,t_1)$,

$$\begin{aligned}|\psi(t')-\phi(t')| &\le \epsilon + K\int_0^{t'}|\psi(s')-\phi(s')|ds \\ &\le \epsilon e^{Kt_1} \le 2\epsilon\end{aligned}$$

by Gronwall's lemma.

Repeating the same argument n_1 times on each $[t_i,t_{i+1})$ we can choose δ,ρ suitably such that

$$\sup_{0\le s\le t}|x^{\eta}_{\ll^{\rho}_{s}}(x)-\phi(s)|<\epsilon$$

holds on Ω_1. □

References

[1] K. Bichteler, J.-B. Gravereaux and J. Jacod, Malliavin Calculus for Processes with Jumps, Gordon and Breach Science Publishers, New York, 1987.

[2] K. Bichteler and J. Jacod, Calcul de Malliavin pour les diffusions avec sauts : existence d'une densite dans le case unidimensionel, in Séminaire de Probabilités XVII (J. Azéma and M. Yor, eds.), Lecture Notes in Math. **986**, Springer-Verlag, Berlin, 1983, 132–157.

[3] J.-M. Bismut, Calcul des variations stochastiques et processus de sauts, Z. Wahrsch. Verw. Gebiete **63** (1983), 147–235.

[4] L. Denis, A criterion of density for solutions of Poisson driven SDE, Preprint, Université du Maine, 2000.

[5] E. Carlen and E. Pardoux, Differential calculus and integration by parts on Poisson space, in Algebra and Analysis in Classical and Quantum dynamics, (S. Albeverio et al. eds.), Kluwer, 1990, 63–73.

[6] T. Fujiwara and H. Kunita, Stochastic differential equations for jump type and Lévy processes in diffeomorphisms group, J. Math. Kyoto Univ. **25** (1985), 71–106.

[7] W. Hoh and N. Jacob, Some Dirichlet forms generated by pseudo differential operators, Bull. Sc. Math. **116** (1992), 383–398.

[8] N. Ikeda and S. Watanabe, Stochastic differential equations and diffusion processes, North-Holland/ Kodansha, Tokyo, 1981.

[9] Y. Ishikawa, On the lower bound of the density for jump processes in small time, Bull. Sc. Math. **117** (1993), 463–483.

[10] Y. Ishikawa, Asymptotic behavior of the transition density for jump-type processes in small time, Tohoku Math. J. **46** (1994), 443–456.

[11] Y. Ishikawa, Large deviation estimate of transition densities for jump processes, Ann. Inst.H.Poincaré Probabilités **33** (1997), 179–222.

[12] Y. Ishikawa, On the upper bound of the density for truncated stable processes in small time, Potential Analysis **6** (1997), 11–37.

[13] Y. Ishikawa, Density estimate in small time for jump processes with singular Lévy measures, Tohoku Math. J., to appear.

[14] J.Jacod, Calcul stochastique et problèm des martingales, Lect. Notes in Math. Vol.714, Springer-Verlag, Berlin, 1975.

[15] H. Kunita, Stochastic differential equations with jumps and stochastic flows of diffeomorphisms, in Itô's stochastic calculus and probability theory (N. Ikeda, S. Watanabe, M. Fukushima and H. Kunita, eds.), Springer-Verlag, Tokyo-Berlin, 1996, 197–211.

[16] R. Léandre, Densité en temps petit d'un processus de sauts, in Séminaire de Probabilités XXI (J. Azéma, P.A.Meyer and M. Yor, eds.), Lecture Notes in Math. 1247, Springer-Verlag, Berlin, 1987, 81–99.

[17] R. Léandre, Régularité de processus de sauts dégénérés (II), Ann. Inst.H.Poincaré Probabilités **24** (1988), 209–236.

[18] Z-M. Ma, M. Röckner and T-S. Zhang, Approximation of arbitrary Dirichlet processes by Markov chains, Ann. Inst. H. Poincaré Probabilités **34** (1998), 1–22.

[19] J.P. Lepeltier and R. Marchal, Problèmes de martingales associées à un opérateur intégro-différentiel, Ann. Inst.H.Poincaré Probabilités **12** (1976), 43–103.

[20] J. R. Norris, Integration by parts for jump processes, in Séminaire de Probabilités XXII (J. Azéma, P.A.Meyer and M. Yor, eds.), Lecture Notes in Math. 1321, Springer-Verlag, Berlin, 1988, 271–315.

[21] J.Picard, On the existence of smooth densities for jump processes, Probab. Th. Relat. Fields **105** (1996), 481–511.

[22] J.Picard, Density in small time for Lévy processes, ESAIM Probab. Statist. **1** (1997), 358–389 (electronic).

[23] J.Picard, Density in small time at accessible points for jump processes, Stochastic Process. Appl. **67** (1997), 251–279.

[24] N. Privault, Linear Skorohod stochastic differential equations on poisson space, in Stochastic Analysis and Related Topics V (H. Körezlioglu, B. Oksendal and A. S. Üstünel, eds.), Birkhäuser, Boston-Basel- Berlin, 1996, 237–253.

[25] P. Protter, Stochastic Integration and Differential Equations, Springer-Verlag, Berlin, 1990.

[26] Th. Simon, Support theorem for jump processes, Stochastic. Process. Appl. **89** (2000), 1–30.

[27] M. Tsuchiya, Lévy measure with generalized polar decomposition and the associated SDE with jumps, Stochastics and Stochastic Reports **38** (1992), 95–117.

Department of Mathematics
Faculty of Science
Ehime University
Matsuyama Ehime 7908571
Japan

Progress in Probability, Vol. 53, 151–167

Controllability and Observability of Linear Stochastic Systems in Hilbert Spaces

Nazim Mahmudov

Abstract. The classical theory of controllability and observability for deterministic systems is extended to linear stochastic time-varying systems defined on infinite dimensional Hilbert spaces. Two types of stochastic controllability (observability) are studied: approximate and complete controllability (observability). Tests for complete and approximate controllability (observability) are proved, and the relation between the controllability (observability) of linear stochastic systems and the controllability (observability) of the corresponding deterministic systems is studied.

1. Notation

In the last few decades several works appeared concerning various kinds of controllability of linear continuous time dynamical systems defined in infinite dimensional spaces. The most popular class of infinite dimensional systems are the systems governed by partial differential equations. In order to investigate the controllability of such systems, we use the theorems concerning infinite dimensional dynamical systems described by abstract differential equations (see [1], [2], [3], and literature therein). Afterwards, many works appeared concerning controllability of stochastic systems. We mention a few here ([4]–[13]).

The present paper systematically studies the notions of controllability for stochastic dynamical systems, described by linear time-varying stochastic differential equations in Hilbert spaces.

In a stochastic setting, there are at least three concepts of controllability; complete controllability, approximate controllability and S-controllability. One of these three concepts, S-controllability of partially observed linear stochastic systems, was studied by Bashirov and this author (see [9]–[11]). The complete and approximate controllabilities of linear time-invariant stochastic systems have been studied in [13].

In Section 2, we give the definition of the completely and approximately controllable partially observable systems. It is known that criteria for controllability in infinite dimensional systems can be obtained by considering the operator $L_0^T(u) = \int_0^T \mathcal{E}(T,t)\,Bu(t)\,dt$ and studying positiveness of $L_0^T(L_0^T)^* = \mathcal{G}_0^T$. To investigate controllability concepts, we define the stochastic analogue of L_0^T, $(L_0^T)^*$, and study the relationship between the controllability map $\mathcal{G}_0^T$ and its stochastic analogue $\mathcal{P}_0^T$. We also find a formula for the control which transfers the linear stochastic system from an arbitrary x_0 to an arbitrary h under natural conditions.

In Sections 3 and 4, with reference to infinite dimensional Hilbert spaces, two types of stochastic controllability are studied; approximate and complete, either of which generalize the classical ones of deterministic systems in infinite dimensional spaces. Our analysis then shows the following:

(i) The complete controllability of the linear stochastic system on $[0,T]$, the complete controllability of the linear stochastic system on every $[s,T]$, and the complete controllability of the corresponding deterministic system are equivalent on every $[s,T]$.

(ii) The approximate controllability of the linear stochastic system on $[0,T]$, the approximate controllability of the linear stochastic system on every $[s,T]$, and the controllability of the corresponding deterministic system on every $[s,T]$ are equivalent.

(iii) The stochastic system is completely (approximately) observable if and only if its dual is completely (approximately) controllable.

- Let $(\Omega, \mathcal{F}, \mathbf{P})$ be a probability space with filtration $\{\mathcal{F}_t : 0 \le t \le T\}$.
- X, H, E and U are separable Hilbert spaces. Let $w_1(t)$ be a Wiener process on E with covariance operator Q_1, and let ξ be an H -valued Gaussian variable with mean x_0 and covariance operator P_0. Let $w_2(t)$ be a vector valued Wiener process on R^k with covariance matrix Q_2. We assume that ξ, w_1, w_2 are mutually independent.
- Let $\mathcal{L}_2\left(Q_1^{1/2}E, H\right)$ be the space of all Hilbert–Schmidt operators from $Q^{1/2}E$ into H endowed with the Hilbert–Schmidt norm $\|\bullet\|_2$.
- $\mathcal{L}(U,H)$ is the space of all linear bounded operators from U to H.
- Let $L_2^{\mathcal{F}}(0,T;X)$ be the space of all $\mathcal{F}_t$-adapted, X-valued measurable processes $\varphi(t,\omega)$ on $[0,T]$ such that $\mathbf{E}\int_0^T \|\varphi(t,\omega)\|^2\,dt < +\infty$.
- $\mathcal{B}(0,T;\mathcal{L}(E,H)) := \{F : [0,T] \to \mathcal{L}(E,H) \mid$

$$\langle x, F(\cdot)\,y\rangle \text{ is measurable for every } x \in H \text{ and } y \in E\}.$$

- $\mathcal{B}_p(0,T;\mathcal{L}(E,H)) := \{F \in \mathcal{B}(0,T;\mathcal{L}(E,H)) \mid$

$$\|F\|_p := \left(\int_0^T \|F(t)\|^p\,dt\right)^{1/p} < \infty\Big\};\ 1 \le p < \infty.$$

- $\mathcal{B}_\infty(0,T;\mathcal{L}(E,H)) := \{F \in \mathcal{B}(0,T;\mathcal{L}(E,H)) \mid \|F\|_\infty := ess\sup_{[0,T]} \|F(t)\| < \infty\}.$
- $L_2(\mathcal{F}_T, H) := L_2(\Omega, \mathcal{F}_T, H).$
- $D(R)$ is a domain and ImR is an image of a linear operator R. A self-adjoint operator R on a Hilbert space H is nonnegative ($R \geq 0$) if $\langle Rz, z\rangle \geq 0$ for all $z \in D(R)$; R is positive ($R > 0$) if $\langle Rz, z\rangle > 0$ for all nonzero $z \in D(R)$; R is coercive ($R - \gamma I \geq 0$) if there exists a $\gamma > 0$ such that $\langle Rz, z\rangle \geq \gamma \|z\|^2$.
- Let f be a $\mathbf{P}$-measurable function on Ω to $L_2(0,T;Y)$. For $0 < t \leq T$, denote by f_t the restriction of f to $[0,t] \times \Omega$, and by $\sigma(f_t)$ denote the σ-algebra generated by $f_t(s), 0 \leq s \leq t$. Let $X_{f_t} = L_2(\Omega, \sigma(f_t), \mathbf{P}, X)$. The set
$$\int^{\oplus} X_{f_t} dt = \{x \in L_2(0,T;X) : x_t \in X_{f_t} \text{ for a.e.} t \in [0,T]\}$$
is the Hilbertian sum of subspaces X_{f_t}.
- Denote $\Delta = \{(t,s) : 0 \leq s \leq t \leq T\}$. Then $\mathcal{E}(\cdot,\cdot) : \Delta \to \mathcal{L}(H)$ is a mild evolution operator if
  1. $\mathcal{E}(t,r)\mathcal{E}(r,s) = \mathcal{E}(t,s)$ for $0 \leq s \leq r \leq t \leq T$, $\mathcal{E}(s,s) = I$,
  2. $\mathcal{E}(t,s)$ is weakly continuous for $0 \leq s \leq t \leq T$.

2. Preliminaries and Definitions

2.1. Partially observable systems

We consider the following class of infinite-dimensional partially observable stochastic systems in the interval $[0,T]$:

$$x(t) = \mathcal{E}(t,0)\xi + \int_0^t \mathcal{E}(t,s)B(s)u(s)ds + \int_0^t \mathcal{E}(t,s)\Sigma(s)dw_1(s), \tag{2.1}$$
$$y(t) = \int_0^t C(s)x(s)ds + \int_0^t F(s)dw_2(s),$$

where $\mathcal{E}(t,s)$ is a mild evolution operator on H, $B \in \mathcal{B}_\infty(0,T;\mathcal{L}(U,H))$, $\Sigma \in \mathcal{B}_2(0,T;\mathcal{L}(E,H))$, $C \in \mathcal{B}_\infty(0,T;\mathcal{L}(H,R^k))$, Q_1 and F are invertible, $F, F^{-1} \in L_2(0,T;\mathcal{L}(R^k))$. Following Bensoussan and Viot [14], in order to avoid the problem of dependence of $\mathcal{U}_{y_t}$ on control process, we take the class of admissible controls

$$\mathcal{U}_{ad} = \int^{\oplus} \mathcal{U}_{y_t^0} dt \bigcap \int^{\oplus} \mathcal{U}_{y_t} dt \subset L_2([0,T] \times \Omega; U),$$

where $y^0(\cdot)$ is the observation process under zero control. For these admissible controls, $u \in \mathcal{U}_{ad}$, it was shown in [14] that $\mathcal{U}_{y_t^0} = \mathcal{U}_{y_t}$ and so is independent of the particular control chosen. So for $u \in \mathcal{U}_{ad}$, (2.1) is a well-defined stochastic process and u is actually "feedback" in the sense that feedback control laws $u(t) = \psi(t, y_t)$, where ψ is measurable, nonanticipative and satisfies a uniform Lipschitz condition, are admissible.

Let $I(t) = \left[y^0(t) - \int_0^t C(s)\, z(s)\, ds\right]$ be an innovation process and let $\mathcal{Y}_t = \sigma(y_t)$, $\mathcal{Y}_t^0 = \sigma(y_t^0)$ and $\mathcal{I}_t = \sigma(I_t)$ be σ-algebras generated by $y_t(\cdot)$, $y_t^0(\cdot)$ and $I_t(\cdot)$, respectively. Introduce an important class of admissible controls, denoted by $\mathcal{U}_{ad}^g$, defined by a linear feedback on the observation:

$$\mathcal{U}_{ad}^g = \left\{ \begin{array}{l} u(t) = \int_0^t K(t,s)\, dy(s) + v(t) : \\ v(\cdot) \in L_2([0,T];U),\ K(\cdot,\cdot) \in \mathcal{B}_2\left(\Delta; \mathcal{L}\left(R^k, U\right)\right) \end{array} \right\}.$$

Lemma 2.1. [15] *$\mathcal{U}_{ad}^g \subset \mathcal{U}_{ad}$. In other words there exist $v_1(\cdot) \in L_2([0,T];U)$ and $R(\cdot,\cdot) \in \mathcal{B}_2\left(\Delta; \mathcal{L}\left(R^k, U\right)\right)$ such that*

$$u(t) = \int_0^t R(t,s)\, dw(s) + v_1(t).$$

The innovation process $I(t)$ is a Wiener process relative to $\mathcal{Y}_t^0$ and $w(t) = F^{-1} I(t)$, where $w(t)$ is a k-dimensional Wiener process with covariance matrix Q.

Lemma 2.2. [16] *If $u(\cdot) \in \mathcal{U}_{ad}$, then there is a unique Kalman filter*

$$z(t) = \mathbf{E}\{x(t) \mid \mathcal{Y}_t\} = \mathbf{E}\{x(t) \mid \mathcal{Y}_t^0\}$$

given by

$$z(t) = \mathcal{E}(t,0)\, x_0 + \int_0^t \mathcal{E}(t,s)\, B(s)\, u(s)\, ds + \int_0^t \mathcal{E}(t,s)\, D(s)\, dw(s), \qquad (2.2)$$

where $D(t) = P(t)\, C^*(t) \left(F(t)\, Q_1 F^*(t)\right)^{-1}$ and $P(t)$ is a unique solution of the Riccati equation

$$\begin{aligned} P(t)\, h = {} & \mathcal{G}(t,0)\, P_0 \mathcal{G}^*(t,0) \\ & + \int_0^t \mathcal{G}(t,s)\, P(s)\, C^*(s) \left(F(s)\, Q_1 F^*(s)\right)^{-1} C(s)\, P(s)\, \mathcal{G}^*(t,s)\, h ds \end{aligned}$$

in the class of weakly continuous self-adjoint operator functions on H. Here $\mathcal{G}(t,s)$ is the perturbation of the mild operator $\mathcal{E}(t,s)$ corresponding to $-D(t)\, C(t)$.

We also consider the deterministic system corresponding to (2.2)

$$y(t) = \mathcal{E}(t,0)\, y_0 + \int_0^t \mathcal{E}(t,s)\, B(s)\, v(s)\, ds, \qquad (2.3)$$

where $v(\cdot) \in L_2(0,T;U)$.

2.2. The stochastic controllability operator

Define the controllability operator $L : L_2^{\mathcal{Y}}(0,T;H) \to L_2(\mathcal{Y}_T, H)$, the controllability grammian $\mathcal{P}_s^T : L_2(\mathcal{Y}_T, H) \to L_2(\mathcal{Y}_T, H)$ associated with (2.2), the controllability grammian $\mathcal{G}_s^T$ associated with (2.3), and $\mathcal{R}(t, x_0)$ the set of all attainable points from x_0 at time t :

$$L_0^T u = \int_0^T \mathcal{E}(T,s) B(s) u(s) \, ds, \tag{2.4}$$

$$\mathcal{P}_s^T = \int_s^T \mathcal{E}(T,t) B(t) B^*(t) \mathcal{E}^*(T,t) \mathbf{E}\{\cdot \mid \mathcal{Y}_t\} dt, \tag{2.5}$$

$$\mathcal{G}_s^T = \int_s^T \mathcal{E}(T,t) B(t) B^*(t) \mathcal{E}^*(T,t) \, dt, \tag{2.6}$$

$$\mathcal{R}(t, x_0) = \{z(t; x_0, u) \mid u \in \mathcal{U}_{ad}\}. \tag{2.7}$$

It is straightforward that operators L_0^T, $\mathcal{P}_s^T$, $\mathcal{G}_s^T$ are linear bounded operators, and the adjoint $\left(L_0^T\right)^* \in \mathcal{L}(L_2(\mathcal{Y}_T, H), \mathcal{U})$ of L_0^T is defined by

$$\left(L_0^T\right)^* z = B^*(t) \mathcal{E}^*(T,t) \mathbf{E}\{z \mid \mathcal{Y}_t\},$$

and

$$\mathcal{P}_0^T = L_0^T \left(L_0^T\right)^*.$$

Before studying the stochastic control problem, let us first investigate the relationship between $\mathcal{P}_s^T$ and $\mathcal{G}_r^T$, $s \le r < T$, and $R\left(\lambda, \mathcal{P}_s^T\right) = \left(\lambda I + \mathcal{P}_s^T\right)^{-1}$ and $R\left(\lambda, \mathcal{G}_r^T\right) = \left(\lambda I + \mathcal{G}_r^T\right)^{-1}$, $s \le r < T$, for $\lambda > 0$, respectively.

Lemma 2.3. *For every $h \in L_2(\mathcal{Y}_T, H)$ there exists $h(\cdot) \in L_2^{\mathcal{Y}}\left(0, T; \ \mathcal{L}\left(R^k, H\right)\right)$ such that*

(a) $h = \mathbf{E}h + \int_0^T h(s) \, dw(s)$,
(b) $\mathcal{P}_s^T h = \mathcal{G}_s^T \mathbf{E}h + \int_s^T \mathcal{G}_r^T h(r) \, dw(r)$,
(c) $R\left(\lambda, \mathcal{P}_s^T\right) h = R\left(\lambda, \mathcal{G}_s^T\right) \mathbf{E}\{h \mid \mathcal{Y}_s\} + \int_s^T R\left(\lambda, \mathcal{G}_r^T\right) h(r) \, dw(r)$.

Proof. For part (a) see [13].

(b) Let $h \in L_2(\mathcal{Y}_T, H)$. From part (a) it follows that there exists $h \in L_2^{\mathcal{Y}}(0,T; \mathcal{L}\left(R^k, H\right))$ such that

$$\mathbf{E}\{h \mid \mathcal{Y}_t\} = \mathbf{E}\{h\} + \int_0^t h(s) \, dw(s).$$

The definition of the operator $\mathcal{P}_s^T$ and stochastic Fubini's theorem lead to the desired representation:

$$\begin{aligned}
\mathcal{P}_s^T h &= \int_s^T \mathcal{E}(T,t) B(t) B^*(t) \mathcal{E}^*(T,t) \mathbf{E}\{h \mid \mathcal{Y}_t\} dt \\
&= \int_s^T \mathcal{E}(T,t) B(t) B^*(t) \mathcal{E}^*(T,t) \left[\mathbf{E}\{h \mid \mathcal{Y}_s\} + \int_s^t h(r)\, dw(r) \right] dt \\
&= \mathcal{G}_s^T \mathbf{E}\{h \mid \mathcal{Y}_s\} + \int_s^T \int_r^T \mathcal{E}(T,t) B(t) B^*(t) \mathcal{E}^*(T,t) h(r)\, dt dw(r) \\
&= \mathcal{G}_s^T \mathbf{E}\{h \mid \mathcal{Y}_s\} + \int_s^T \mathcal{G}_r^T h(r)\, dw(r).
\end{aligned}$$

(c) This follows from (a) and (b). □

Now we define the stochastic analogue of complete controllability and approximate controllability concepts.

Definition 2.4. *The linear stochastic system (2.2) is said to be*

(a) *completely controllable on* $[0,T]$ *if all the points in* $L_2(\mathcal{Y}_T, H)$ *can be reached from the point* x_0 *at time* T*, that is, if*

$$\mathcal{R}(T, x_0) = L_2(\mathcal{Y}_T, H).$$

(b) *approximately controllable on* $[0,T]$ *if*

$$\overline{\mathcal{R}(T, x_0)} = L_2(\mathcal{Y}_T, H),$$

i.e., it is possible to steer from the point x_0 *to within a distance* $\varepsilon > 0$ *from all points in the state space* $L_2(\mathcal{Y}_T, H)$ *at time* T.

If $T > 0$ can be arbitrarily small, we add the words "small time" in front of "controllable." We say small time exactly controllable, small time approximately controllable and small time S-controllable.

2.3. Minimum energy principle

Define the linear regulator problem: to minimize

$$J(u) = \mathbf{E}\| z(T; x_0, u) - h\|^2 + \lambda \mathbf{E} \int_0^T \|u(t)\|^2 dt, \tag{2.8}$$

over all $u(\cdot) \in \mathcal{U}_{ad}$, where $z(t; x_0, u)$ is a state process, defined by (2.2), and $h \in L_2(\mathcal{Y}_T, H)$ and $\lambda > 0$ are parameters.

Lemma 2.5. *There exists a unique optimal control* $u^\lambda(\cdot) \in \mathcal{U}_{ad}$, *at which the functional (2.8) takes on its minimum value and*

$$u^\lambda(t) = -B^*(t)\,\mathcal{E}^*(T,t)\left\{R(\lambda,\mathcal{G}_0^T)(\mathcal{E}(T,0)\,x_0 - \mathbf{E}h) + \int_0^t R\left(\lambda,\mathcal{G}_s^T\right)[\mathcal{E}(T,s)\,D(s) - h(s)]\,dw(s)\right\}, \tag{2.9}$$

$$z\left(T;x_0,u^\lambda\right) - h = \lambda R(\lambda,\mathcal{G}_0^T)(\mathcal{E}(T,0)\,x_0 - \mathbf{E}h) + \int_0^T \lambda R\left(\lambda,\mathcal{G}_s^T\right)[\mathcal{E}(T,s)\,D(s) - h(s)]\,dw(s), \tag{2.10}$$

where $h(\cdot) \in L_2^{\mathcal{Y}}\left(0,T;\mathcal{L}\left(R^k,H\right)\right)$ *can be found from the representation*

$$h = \mathbf{E}h + \int_0^T h(s)\,dw(s).$$

Proof. The problem of minimizing the functional (2.8) has a unique solution $u^\lambda(\cdot) \in \mathcal{U}_{ad}$ which is completely characterized by the stochastic maximum principle and has the form

$$u^\lambda(t) = -\lambda^{-1}B^*(t)\,\mathcal{E}^*(T,t)\,\mathbf{E}\left\{z\left(T;x_0,u^\lambda\right) - h \mid \mathcal{Y}_t\right\}. \tag{2.11}$$

Using this in (2.2), we have

$$\begin{aligned}
\hat{x}\left(T;x_0,u^\lambda\right) &= \mathcal{E}(T,0)\,x_0 + \int_0^T \mathcal{E}(T,t)\,B(t)\,u^\lambda(t)\,dt \\
&+ \int_0^T \mathcal{E}(T,t)\,D(t)\,dw(t) \\
&= \mathcal{E}(T,0)\,x_0 + \int_0^T \mathcal{E}(T,t)\,D(t)\,dw(t) \\
&- \lambda^{-1}\int_0^T \mathcal{E}(T,t)\,B(t)\,B^*(t)\,\mathcal{E}^*(T,t)\,\mathbf{E}\left\{z\left(T;x_0,u^\lambda\right) - h \mid \mathcal{Y}_t\right\}dt \\
&= \mathcal{E}(T,0)\,x_0 + \int_0^T \mathcal{E}(T,t)\,D(t)\,dw(t) \\
&- \lambda^{-1}\mathcal{P}_0^T\left(z\left(T;x_0,u^\lambda\right) - h\right).
\end{aligned}$$

Hence,

$$\lambda z\left(T;x_0,u^\lambda\right) = \lambda\left(\mathcal{E}(T,0)\,x_0 + \int_0^T \mathcal{E}(T,t)\,D(t)\,dw(t)\right) - \mathcal{P}_0^T\left(z\left(T;x_0,u^\lambda\right) - h\right),$$

which implies

$$\left(\lambda I+\mathcal{P}_0^T\right) z\left(T; x_0, u^\lambda\right)=\lambda\left(\mathcal{E}(T, 0) x_0+\int_0^T \mathcal{E}(T, t) D(t) d w(t)\right)+\mathcal{P}_0^T h,$$

and consequently, from Lemma 2.3 we obtain:

$$\begin{aligned} z\left(T; x_0, u^\lambda\right)-h= & \lambda(\lambda I+\mathcal{P}_0^T)^{-1}\left(\mathcal{E}(T, 0) x_0+\int_0^T \mathcal{E}(T, t) D(t) d w(t)\right) \\ & +(\lambda I+\mathcal{P}_0^T)^{-1}(\lambda I+\mathcal{P}_0^T-\lambda I) h-h \\ = & \lambda R(\lambda, \mathcal{P}_0^T)\left(\mathcal{E}(T, 0) x_0+\int_0^T \mathcal{E}(T, t) D(t) d w(t)-h\right) \\ = & \lambda R(\lambda, \mathcal{G}_0^T)(\mathcal{E}(T, 0) x_0-\mathbf{E} h) \\ & +\int_0^T \lambda R(\lambda, \mathcal{G}_s^T)[\mathcal{E}(T, t) D(t)-h(t)] d w(t) . \end{aligned}$$

Thus, (2.10) holds. Substituting (2.10) in (2.11), we obtain (2.9). The lemma is proved. □

In the case when a linear system (2.2) is completely controllable in the interval $[0, T]$, there exist in general many different controls $u(\cdot)$ that can steer the initial state x_0 to the desired final state $h \in L_2(\mathcal{Y}_T, H)$ at time T. We may ask which of these possible controls is the optimal one according to some criterion given a priori. The next lemma gives a formula for a control transferring x_0 to an arbitrary h, which minimizes the integral $I(u)=\mathbf{E} \int_0^T \|u(t)\|^2 dt$ (for the deterministic analogue of this formula see [1], [3]).

Lemma 2.6. *Assume that for arbitrary $0 \leq t < T$ the operator $\mathcal{G}_t^T$ is invertible. Then*

(i) *for arbitrary $x_0 \in H$ and $h \in L_2(\mathcal{Y}_T, H)$ the control*

$$\begin{aligned} u^0(t)= & B^*(t) \mathcal{E}^*(T, t)\left\{\left(\mathcal{G}_0^T\right)^{-1}(\mathbf{E} h-\mathcal{E}(T, 0) x_0)\right. \\ & \left.+\int_0^t\left(\mathcal{G}_s^T\right)^{-1}[h(s)-\mathcal{E}(T, s) D(s)] d w(s)\right\} \end{aligned} \tag{2.12}$$

transfers x_0 to h at time T, where $h(\cdot) \in L_2^{\mathcal{Y}}\left(0, T ; \mathcal{L}\left(R^k, H\right)\right)$ can be found from the representation

$$h=\mathbf{E} h+\int_0^T h(s) d w(s);$$

(ii) *among all controls u transferring x_0 to h at time T the control $u^0(\cdot)$ minimizes the integral $I(u)=\mathbf{E} \int_0^T \|u(t)\|^2 dt$.*

Proof. (i) We will show that the control $u^0(\cdot)$ given by formula (2.12) steers the stochastic system (2.2) from the initial state x_0 to the desired final state h at time T. Substituting (2.12) into (2.2), we obtain for T the equality

$$
\begin{aligned}
z\left(T;x_0,u^0\right) &= \mathcal{E}(T,0)\,x_0 + \int_0^T \mathcal{E}(T,t)\,B(t)\,u^0(t)\,ds \\
&\quad + \int_0^T \mathcal{E}(T,t)\,D(t)\,dw(t) \\
&= \mathcal{E}(T,0)\,x_0 + \int_0^T \mathcal{E}(T,t)\,D(t)\,dw(t) \\
&\quad + \int_0^T \mathcal{E}(T,t)\,B(t)\,B^*(t)\,\mathcal{E}^*(T,t) \\
&\quad \times \left\{\left(\mathcal{G}_0^T\right)^{-1}\left(\mathbf{E}h - \mathcal{E}(T,0)\,x_0\right)\right. \\
&\quad \left. + \int_0^t \left(\mathcal{G}_s^T\right)^{-1}\left[h(s) - \mathcal{E}(T,s)\,D(s)\right]dw(s)\right\}dt \\
&= \mathcal{E}(T,0)\,x_0 + \int_0^T \mathcal{E}(T,t)\,D(t)\,dw(t) + \mathbf{E}h - \mathcal{E}(T,0)\,x_0 \\
&\quad + \int_0^T \int_s^T \mathcal{E}(T,t)\,B(t)\,B^*(t)\,\mathcal{E}^*(T,t)\left(\mathcal{G}_s^T\right)^{-1}\left[h(s) - \mathcal{E}(T,s)\,D(s)\right]dt\,dw(s) \\
&= \int_0^T \mathcal{E}(T,t)\,D(t)\,dw(t) + \mathbf{E}h + \int_0^T \left[h(s) - \mathcal{E}(T,s)\,D(s)\right]dw(s) \\
&= \mathbf{E}h + \int_0^T h(s)\,dw(s) = h.
\end{aligned}
$$

From the above equality it follows that the control $u^0(\cdot)$ steers the system (2.2) from x_0 to h.

(ii) Suppose the control $u'(\cdot) \in \mathcal{U}_{ad}$ is any control which steers the system (2.2) from x_0 to h at time T. Hence we have

$$
\int_0^T \mathcal{E}(T,t)\,B(t)\,u'(t)\,dt = \int_0^T \mathcal{E}(T,t)\,B(t)\,u^0(t)\,dt,
$$

$$
\int_0^T \mathcal{E}(T,t)\,B(t)\left(u'(t) - u^0(t)\right)dt = 0,
$$

which implies that

$$\mathbf{E}\left\langle \int_0^T \mathcal{E}(T,t)\,B(t)\left(u'(t)-u^0(t)\right)dt, \left(\mathcal{G}_0^T\right)^{-1}\left(\mathbf{E}h-\mathcal{E}(T,0)\,x_0\right) + \int_0^T \left(\mathcal{G}_s^T\right)^{-1}\left[h(s)-\mathcal{E}(T,s)\,D(s)\right]dw(s)\right\rangle$$

$$= \mathbf{E}\int_0^T \left\langle u'(t)-u^0(t), \mathcal{E}^*(T,t)\,B^*(t)\left\{\left(\mathcal{G}_0^T\right)^{-1}\left(\mathbf{E}h-\mathcal{E}(T,0)\,x_0\right) + \int_0^T \left(\mathcal{G}_s^T\right)^{-1}\left[h(s)-\mathcal{E}(T,s)\,D(s)\right]dw(s)\right\}\right\rangle dt$$

$$= \mathbf{E}\int_0^T \left\langle u'(t)-u^0(t), u^0(t) + \mathcal{E}^*(T,t)\,B^*(t)\int_t^T \left(\mathcal{G}_s^T\right)^{-1}\left[h(s)-\mathcal{E}(T,s)\,D(s)\right]dw(s)\right\rangle dt,$$

$$\mathbf{E}\int_0^T \left\langle u'(t)-u^0(t), u^0(t)\right\rangle dt = 0.$$

Using the properties of the scalar product and taking into account the last formula we have

$$\begin{aligned}\mathbf{E}\int_0^T \left\|u'(t)\right\|^2 dt &= \mathbf{E}\int_0^T \left\langle u'(t)-u^0(t)+u^0(t), u'(t)-u^0(t)+u^0(t)\right\rangle dt\\ &= \mathbf{E}\int_0^T \left\|u'(t)-u^0(t)\right\|^2 dt + \mathbf{E}\int_0^T \left\|u^0(t)\right\|^2 dt\\ &\geq \mathbf{E}\int_0^T \left\|u^0(t)\right\|^2 dt.\end{aligned}$$

Hence we have proved part (ii), which states that the control $u^0(\cdot)$ is the optimal one for the functional $I(u)$. □

3. Controllability Conditions for Linear Systems

3.1. The Complete Controllability

In this section necessary and sufficient conditions for complete controllability are discussed.

Theorem 3.1. *Let A be a self-adjoint, and nonnegative operator on X. Then the following three conditions are equivalent.*

(i) *A is coercive, that is, $\langle Ax, x\rangle \geq \gamma \|x\|^2$ for all $x \in D(A)$.*
(ii) *$R(\lambda, A)$ converges as $\lambda \to 0^+$ in the uniform operator topology.*

(iii) $\lambda R(\lambda, A)$ *converges to the zero operator as* $\lambda \to 0^+$ *in the uniform operator topology.*

Proof. (i) $\Rightarrow$ (ii) Suppose $\langle Ax, x\rangle \geq \gamma \|x\|^2$ for all $x \in D(A)$. Then for all $x \in D(A)$ and for all $\lambda \geq 0$,

$$\langle x, (\lambda I + A)x\rangle \geq (\lambda + \gamma)\|x\|^2. \tag{3.1}$$

From (3.1), we see that $(\lambda I + A)$ is injective. So $R(\lambda, A) = (\lambda I + A)^{-1} : \operatorname{Im}(\lambda I + A) \to X$ is well defined. For all $z \in \operatorname{Im}(\lambda I + A)$ we have

$$\|(\lambda I + A)^{-1} z\| \leq \frac{1}{\lambda + \gamma}\|z\| \leq \frac{1}{\gamma}\|z\| \tag{3.2}$$

which means that $(\lambda I + A)^{-1}$ is bounded on $\operatorname{Im}(\lambda I + A)$. So, $\operatorname{Im}(\lambda I + A)$ is closed. Now, let us show that $D(R(\lambda, A)) = \operatorname{Im}(\lambda I + A)$ is dense in H. Let $x \in [\operatorname{Im}(\lambda I + A)]^{\perp} = N(\lambda I + A^*)$. Since $(\lambda I + A)$ is self-adjoint, we conclude that

$$[\operatorname{Im}(\lambda I + A)]^{\perp} = N(\lambda I + A^*) = N(\lambda I + A),$$

and since $(\lambda I + A)$ is coercive, then $N(\lambda I + A) = \{0\}$. So $\operatorname{Im}(\lambda I + A)$ is dense in H. Thus, $(\lambda I + A)^{-1}$ is bounded on a closed and dense set $\operatorname{Im}(\lambda I + A)$. It means that $\operatorname{Im}(\lambda I + A) = H$ and $R(\lambda, A) = (\lambda I + A)^{-1} \in \mathcal{L}(X)$ is well defined, and by the formula (3.2)

$$\|R(\lambda, A)\| = \|(\lambda I + A)^{-1}\| \leq \frac{1}{\gamma},$$

that is, $\|R(\lambda, A)\|$ is uniformly bounded with respect to $\lambda \geq 0$. Furthermore,

$$\begin{aligned}\|R(\lambda, A) - A^{-1}\| &= \|(\lambda I + A)^{-1} - A^{-1}\| \\ &= \|A^{-1}(A - \lambda I - A)(\lambda I + A)^{-1}\| \\ &\leq \lambda\|A^{-1}\|\,\|(\lambda I + A)^{-1}\| \leq \frac{\lambda}{\gamma^2}.\end{aligned}$$

So, $R(\lambda, A)$ converges to A^{-1} as $\lambda \to 0^+$ in the uniform operator topology.

(ii) $\Rightarrow$ (iii) This implication is obvious.

(iii) $\Rightarrow$ (i) Suppose

$$\lambda\|R(\lambda, A)\| = \lambda\left\|(\lambda I + A)^{-1}\right\| \to 0, \ \lambda \to 0^+.$$

Then $\lambda^{1/2}\left\|(\lambda I + A)^{-1/2}\right\| \to 0$ as $\lambda \to 0^+$. For sufficiently small $\gamma > 0$, we can write

$$\gamma^{1/2}\left\|(\gamma I + A)^{-1/2}\right\| \leq 1/\sqrt{2}.$$

So, for all $x \in X$, we have

$$\begin{aligned}\|x\|^2 &= \left\|\left(\gamma^{1/2}(\gamma I + A)^{-1/2}\right)\left(\gamma^{-1/2}(\gamma I + A)^{1/2}\right)x\right\|^2 \\ &\leq \frac{1}{2}\left\|\gamma^{-1/2}(\gamma I + A)^{1/2}x\right\|^2 \\ &= \frac{1}{2}\left\langle \gamma^{-1}(\gamma I + A)x, x\right\rangle,\end{aligned}$$

which implies

$$\left\langle \gamma^{-1}(\gamma I + A)x, x\right\rangle \geq 2\left\|x\right\|^2 \Leftrightarrow \langle Ax, x\rangle \geq \gamma\left\|x\right\|^2.$$

□

Theorem 3.2. *The control system (2.2) is completely controllable on $[0,T]$ if and only if any one of the following conditions hold:*

1. $\mathbf{E}\left\langle \mathcal{P}_0^T z, z\right\rangle \geq \gamma\mathbf{E}\left\|z\right\|^2$ *for some* $\gamma > 0$ *and all* $z \in L_2(\mathcal{Y}_T, H)$.
2. $R(\lambda, \mathcal{P}_0^T)$ *converges as* $\lambda \to 0^+$ *in the uniform operator topology.*
3. $\lambda R(\lambda, \mathcal{P}_0^T)$ *converges to the zero operator as* $\lambda \to 0^+$ *in the uniform operator topology.*
4. *ker* $\left(L_0^T\right)^* = \{0\}$ *and* $\operatorname{Im}\left(L_0^T\right)^*$ *is closed.*

3.2. The Approximate Controllability

In practice, most dynamical systems defined in infinite dimensional spaces are only approximately controllable and not completely controllable. In this section we study approximately controllable linear stochastic systems.

Theorem 3.3. *Let A be a self-adjoint and nonnegative operator on X. Then the following three conditions are equivalent:*

(i) *A is positive, that is, $\langle Ax, x\rangle > 0$ for all nonzero $x \in D(A)$.*
(ii) *$\lambda R(\lambda, A)$ converges to the zero operator as $\lambda \to 0^+$ in the weak operator topology.*
(iii) *$\lambda R(\lambda, A)$ converges to the zero operator as $\lambda \to 0^+$ in the strong operator topology.*

Proof. (i) ⇒ (ii) Let A be a positive operator, but there exists $y \neq 0$ such that $\lim_{\lambda\to 0^+} \langle \lambda R(\lambda, A)x, z\rangle = \langle y, z\rangle$ for all $z \in X$. Set $\lambda(\lambda I + A)^{-1} x = z^\lambda$. Then

$$\begin{aligned}
\lambda x &= (\lambda I + A)\, z^\lambda, \\
\lambda\langle x, z\rangle &= \lambda\left\langle z^\lambda, z\right\rangle + \left\langle Az^\lambda, z\right\rangle \\
&= \lambda\left\langle z^\lambda, z\right\rangle + \left\langle z^\lambda, Az\right\rangle \text{ for all } z \in D(A) = D(A^*),
\end{aligned}$$

and taking the limit from both sides we obtain

$$\langle y, Az\rangle = \text{ for all } z \in D(A).$$

This implies that $y \in D(A^*)$ and $A^*y = 0$. Since A is self-adjoint, we conclude that $Ay = A^*y = 0$ for nonzero $y \in D(A)$, which contradicts the positivity of A.

(ii) $\Rightarrow$ (iii) Assume that $\langle \lambda R(\lambda, A)x, y\rangle \to 0$ as $\lambda \to 0^+$ for all $x, y \in X$. Since $R(\lambda, A) \in \mathcal{L}(X)$ is nonnegative and self-adjoint, we deduce that

$$\|\lambda R(\lambda, A)x\|^2 \leq \|\lambda R(\lambda, A)\| \langle \lambda R(\lambda, A)x, x\rangle \leq \langle \lambda R(\lambda, A)x, x\rangle \to 0.$$

So $\lambda R(\lambda, A)x \to 0$ in the strong topology.

(iii) $\Rightarrow$ (i) Now, assume that $\lim_{\lambda \to 0^+} \lambda R(\lambda, A)x = 0$, but there exists nonzero $x \in D(A)$ such that $\langle Ax, x\rangle = 0$. Then there exists a unique self-adjoint $\sqrt{A} \geq 0$ such that $D(A) \subset D\left(A^{1/2}\right)$ and

$$\langle Ax, x\rangle = \left\langle \left(\sqrt{A}\right)^2 x, x\right\rangle = \left\|\sqrt{A}x\right\|^2 = 0,$$

which implies that $\sqrt{A}x = 0$ and, consequently $Ax = 0$. Then $x = \lambda\left(\lambda I + A\right)^{-1} x$, and so $\lim_{\lambda \to 0^+} \lambda R(\lambda, A)x \neq 0$ which leads to a contradiction. $\square$

Theorem 3.4. *The control system (2.2) is approximately controllable on $[0,T]$ if and only if any one of the following conditions hold:*

1. $\mathcal{P}_0^T > 0$.
2. $\lambda R(\lambda, \mathcal{P}_0^T)$ *converges to the zero operator as $\lambda \to 0^+$ in the strong operator topology.*
3. $\lambda R(\lambda, \mathcal{P}_0^T)$ *converges to the zero operator as $\lambda \to 0^+$ in the weak operator topology.*
4. $\ker\left(L_0^T\right)^* = \{0\}$.

Theorem 3.5. *The following three conditions are equivalent:*

(a) *The stochastic system (2.2) is approximately controllable on $[0,T]$.*
(b) *The deterministic system (2.3) is approximately controllable on every $[s,T]$, $0 \leq s < T$.*
(c) *The stochastic system (2.2) is approximately controllable on every $[s,T]$, $0 \leq s < T$.*

Proof. (a) $\Rightarrow$ (b) : Let the stochastic system (2.2) be approximately controllable on $[0,T]$. Then

$$\mathbf{E}\left\|\lambda R(\lambda, \mathcal{P}_0^T)h\right\|^2 \to 0.$$

From this and the equation (2.10), we have

$$\mathbf{E}\left\|\lambda R(\lambda, \mathcal{P}_0^T)h\right\|^2 = \left\|\lambda R(\lambda, \mathcal{G}_0^T)\mathbf{E}h\right\|^2$$
$$+\mathbf{E}\sum_{j=1}^{k} \alpha_j \int_0^T \left\|\lambda R\left(\lambda, \mathcal{G}_s^T\right) h_j(s)\right\|^2 ds \to 0. \tag{3.3}$$

From here

$$\mathbf{E}\sum_{j=1}^{k}\alpha_j\int_0^T\left\|\lambda R\left(\lambda,\mathcal{G}_s^T\right)h_j\left(s\right)\right\|^2ds\to 0$$

for all $h\left(\cdot\right)\in L_2^{\mathcal{Y}}(0,T,\mathcal{L}_2\left(R^k,H\right))$, and consequently there is a subsequence $\{\lambda_k\}$ such that, for all $g\in H$,

$$\left\|\lambda_k R\left(\lambda_k,\mathcal{G}_s^T\right)g\right\|\to 0,\text{ almost everywhere on }\left[0,T\right].$$

Because of continuity of $R\left(\lambda,\mathcal{G}_s^T\right)$ this property holds for all $0\leq s<T$. The latter means that the deterministic system (2.3) is approximately controllable on every $\left[s,T\right]$, $0\leq s<T$.

(b) $\Rightarrow$ (c) : If the deterministic system (2.3) is approximately controllable on every $\left[s,T\right]$, then $\left\|\lambda R\left(\lambda,\mathcal{G}_s^T\right)g\right\|\to 0$ as $\lambda\to 0^+$. Since

$$\sum_{j=1}^{k}\alpha_j\left\|\lambda R\left(\lambda,\mathcal{G}_s^T\right)h_j\left(s\right)\right\|^2\leq\sum_{j=1}^{k}\alpha_j\left\|h_j\left(s\right)\right\|^2,$$

by the Lebesgue dominated convergence theorem and from (3.3) we get

$$\begin{aligned}&\mathbf{E}\left\|\lambda R(\lambda,\mathcal{P}_r^T)h\right\|^2\\&=\left\|\lambda R(\lambda,\mathcal{G}_0^T)\mathbf{E}h\right\|^2+\mathbf{E}\sum_{j=1}^{k}\alpha_j\int_r^T\left\|\lambda R\left(\lambda,\mathcal{G}_s^T\right)h_j\left(s\right)\right\|^2ds\to 0\text{ as }\lambda\to 0^+,\end{aligned}$$

that is, the stochastic system (2.2) is approximately controllable on every $\left[r,T\right]$.

(c) $\Rightarrow$ (a) : This implication is obvious. □

Theorem 3.6. *For the time-invariant systems (2.2) and (2.3) the following four conditions are equivalent:*

1. *The stochastic system (2.2) is approximately controllable on* $\left[0,T\right]$.
2. *The deterministic system (2.3) is approximately controllable on every* $\left[s,T\right]$, $0\leq s<T$.
3. *The deterministic system (2.3) is small time approximately controllable.*
4. *The stochastic system (2.2) is small time approximately controllable.*

4. Observability

In this section we consider the dual concepts of observability for linear stochastic time-invariant systems. The idea behind observability is the property that knowledge of the output data over a finite interval uniquely determines the initial state.

We consider the time-invariant system

$$\begin{aligned} dx(t) &= Ax(t)\,dt + Ddw(t), \\ y(t) &= Cx(t), \quad x(0) = x_0, \end{aligned} \tag{4.1}$$

where A generates a C_0-semigroup $\mathcal{S}(\cdot)$ on a Hilbert space H, $C \in \mathcal{L}(H, Y)$ and its dual

$$\begin{aligned} dx(t) &= [A^*x(t) + C^*u(t)]\,dt + Ddw(t), \\ x(0) &= x_0 \end{aligned} \tag{4.2}$$

For the linear stochastic system (4.1), we define the following concepts:

- The observability map of (4.1) on $[0,T]$ is the bounded linear operator $K_0^T : L_2(\mathcal{F}_T, H) \to L_2^{\mathcal{F}}(0,T;Y)$ defined by
$$K_0^T z = C\mathcal{S}(T-\cdot)\,\mathbf{E}\{z \mid \mathcal{F}_\cdot\}.$$
- The observability grammian of (4.1) on $[0,T]$,
$$\mathcal{Q}_0^T = \left(K_0^T\right)^* K_0^T.$$

Definition 4.1. *The linear stochastic system (2.2) is said to be*

(a) *completely observable on $[0,T]$ if K_0^T is injective and its inverse is bounded on the range of K_0^T, i.e., the initial state can be uniquely and continuously constructed from the knowledge of the output in $L_2^{\mathcal{F}}(0,T;Y)$;*

(b) *approximately observable on $[0,T]$ if*
$$\ker K_0^T = \{0\},$$
i.e., knowledge of the output in $L_2^{\mathcal{F}}(0,T;Y)$ determines the initial state uniquely.

Lemma 4.2. *For the linear stochastic system (4.1) we have the following duality results:*

1. *The system (4.1) is completely observable on $[0,T]$ if and only if the dual system (4.2) is completely controllable on $[0,T]$.*
2. *The system (4.1) is approximately observable on $[0,T]$ if and only if the dual system (4.2) is approximately controllable on $[0,T]$.*

Proof. It is known that if A generates a C_0-semigroup $\mathcal{S}(t)$ on a Hilbert space H, then A^* generates the C_0-semigroup $\mathcal{S}^*(t)$. Moreover, since $K_0^T \in \mathcal{L}\big(L_2(\mathcal{F}_T, H), L_2^{\mathcal{F}}(0,T;Y)\big)$ we obtain

$$\left(K_0^T\right)^* y = \int_0^T \mathcal{S}^*(T-t)\,C^*y(t)\,dt,$$

and the range of $\left(K_0^T\right)^*$ equals that of the controllability operator for the dual system (4.2). If we define the controllability operator of this dual system (4.2) by N_0^T, then we have that

$$\left(K_0^T\right)^* = N_0^T \text{ and } \left(N_0^T\right)^* = K_0^T.$$

1. Assume that the system (4.1) is completely observable. So there is an inverse $\left(K_0^T\right)^{-1}$ on the range of K_0^T. Then

$$\left\| \left(K_0^T\right)^{-1} y \right\| \leq l \left\|y\right\| \text{ for all } y \in \operatorname{Im} K_0^T.$$

Hence we have that

$$\left\|z\right\| = \left\| \left(K_0^T\right)^{-1} K_0^T z \right\| \leq l \left\|K_0^T z\right\| = l \left\| \left(N_0^T\right)^* z \right\|.$$

The complete controllability of the system (4.2) now follows from Theorem 3.2.

Suppose next that the system (4.2) is completely controllable. By Theorem 3.2 we have that $\left(N_0^T\right)^*$ is injective and has closed range. From the fact that $\left(N_0^T\right)^* = K_0^T$, this implies that K_0^T is injective and has closed range. By the Closed Graph Theorem the inverse of K_0^T is bounded on the range of K_0^T.

2. By definition, the system (4.1) is approximately observable if and only if

$$\ker K_0^T = \ker \left(N_0^T\right)^* = \{0\}.$$

By Theorem 3.4, $\ker \left(N_0^T\right)^* = \{0\}$ if and only if the system (4.2) is approximately controllable, which proves the equivalence. □

Corollary 4.3. *For the linear stochastic system (2.2), we have the following necessary and sufficient conditions for complete and approximate observability:*

1. *The system (2.2) is completely observable on* $[0,T]$ *if and only if any one of the following conditions holds:*
   (a) $\mathbf{E}\left\langle Q_s^T z, z\right\rangle \geq \gamma \mathbf{E}\left\|z\right\|^2$ *for some* $\gamma > 0$ *and all* $z \in L_2(\mathcal{Y}_T, H)$.
   (b) $R(\lambda, Q_s^T)$ *converges as* $\lambda \to 0^+$ *in the uniform operator topology.*
   (c) $\lambda R(\lambda, Q_s^T)$ *converges to the zero operator as* $\lambda \to 0^+$ *in the uniform operator topology.*
   (d) $\ker K_0^T = \{0\}$ *and* K_0^T *has closed range.*
2. *The system (2.2) is approximately observable on* $[0,T]$ *if and only if any one of the following conditions holds:*
   (a) $Q_0^T > 0$.
   (b) $\lambda R(\lambda, Q_0^T)$ *converges to the zero operator as* $\lambda \to 0^+$ *in the strong operator topology.*
   (c) $\lambda R(\lambda, Q_0^T)$ *converges to the zero operator as* $\lambda \to 0^+$ *in the weak operator topology.*
   (d) $\ker K_0^T = \{0\}$.

The duality can be used to obtain results about observability from ones about controllability (see [13]).

References

[1] A. Bensoussan, G. Da Prato, M. C. Delfour and S. K. Mitter, "Representation and Control of Infinite Dimensional Systems," Volume 2, Systems & Control: Foundations & Applications, Birkhäuser/Boston, 1993.

[2] R. F. Curtain and H. J. Zwart, "An Introduction to Infinite Dimensional Linear Systems Theory," Springer-Verlag/New York, 1995.

[3] J. Zabczyk, "Mathematical Control Theory," Birkhäuser/Boston, Basel, Berlin, 1992.

[4] Y. Sunahara, T. Kabeuchi, S. Asada, S. Aihara and K. Kishino, On Stochastic Controllability for Nonlinear Systems, *IEEE Trans. Automat. Control*, **19** (1974), 49–54.

[5] J. Klamka and L. Socha, Some remarks about stochastic controllability, *IEEE Trans. Automat. Control*, **22** (1977), 880–881.

[6] J. Zabczyk, Controllability of Stochastic Linear Systems, *Syst. Control. Lett.* **1** (1991), 25–31.

[7] M. Ehrhardt and W. Kliemann, Controllability of Stochastic Linear Systems, *Syst. Control. Lett.* **2** (1982), 145–153

[8] M.A. Dubov and B.S. Mordukhovich, On controllability of infinite-dimensional linear stochastic systems, in "IFAC Proceedings Series" Vol. 2, Pergamon Press/Oxford-New York, 1987.

[9] A. E. Bashirov and K. R. Kerimov, On controllability conception for stochastic systems. *SIAM J. Control and Optimization*, **35** (1997), 384–398.

[10] A. E. Bashirov and N. I. Mahmudov, On Concepts of Controllability for Deterministic and Stochastic Systems. *SIAM J. on Control and Optim.* **37** (1999), 1808–1821.

[11] A. E. Bashirov and N. I. Mahmudov, Controllability of Linear Deterministic and Stochastic Systems, *Proceedings of the 38th IEEE Conference on Decision and Control*, Phoenix, Arizona, USA, December 7-10, Vol. 4 (1999), 3196–3202.

[12] N. I. Mahmudov and A.Denker, On Controllability of linear stochastic systems. *Int. J. Control*, **73** (2000), 144–151.

[13] N. I. Mahmudov, Controllability of Linear Stochastic Systems in Hilbert Spaces, *Journal of Mathematical Analysis and Appl.* **259**, (2001), 64–82.

[14] A. Bensoussan and M. Viot, Optimal Control of Stochastic Linear Distributed Parameter systems, *SIAM J. Control*, **13** (1975), 904–926.

[15] A. Lindquist, On feedback control on linear stochastic systems, *SIAM J. Control,* **11** (1973), 323–343.

[16] R. F. Curtain and A. Ichikawa, The Separation Principle for Stochastic Evolution Equations, *SIAM J. on Control and Optim.*, **15** (1977), 367–383.

[17] R. S. Liptster and A. N. Shiryaev, "Statistics of Random Processes," Springer-Verlag/New York, 1977.

Department of Mathematics
Eastern Mediterranean University
G. Magusa
North Cyprus (via Mersin 10 – Turkey)

Progress in Probability, Vol. 53, 169–180

Exact Computation of Spectral Densities for a Langevin Dynamic Modelling the Relaxation of a Protein near its Equilibrium

Paul Malliavin and Ufuk Taneri

Abstract. The relaxation motion of a protein nearby its equilibrium is enhanced by the heat bath in which the protein is embedded. A classical approach will realize this heat bath through a molecular dynamic including a large number of molecules of the solvent at the cost of a large computing time. Here, we propose a modelling of the heat bath by a Langevin random force. Under this Langevin dynamic, an exact mathematical computation of the spectral densities can be made.The mathematical expression will depend upon the normal modes of the potential nearby its minimum. The mathematical computation is exact; it is, therefore, unnecessary to run an effective simulation through a molecular dynamic. The spectral densities will peak at these normal modes. Our computation provides a theoretical ground to the experimental determination of normal modes through a peak searching of observed spectral densities.

1. Hamiltonian System

The Hamiltonian of a system of N particles of masses m_i is given by the expression

$$H(p,q) = \frac{1}{2}\sum_i \frac{1}{m_i} \|p_i\|^2 + V(q),$$

and the Hamilton equations of the motion can be written

$$\dot{p} + \frac{\partial H}{\partial q} = 0, \tag{1.1}$$

$$\dot{q} - \frac{\partial H}{\partial p} = 0. \tag{1.2}$$

Einstein modelled the diffusion of a pollen grain by introducing the "Brownian motion" denoted here by β. The Brownian motion is an R^3-valued motion characterized by the following properties of its variation. Given $I = [t_1, t_2]$, $t_1 < t_2$, a time interval, we denote interval $\delta_I\beta = \beta(t_2) - \beta(t_1)$ and $|I| = t_2 - t_1$.

Then the Brownian motion is characterized by the following two properties:

- for each interval I
$$\frac{1}{\sqrt{|I|}}\delta_I(\beta) \tag{1.3}$$
is a *gaussian normal variable,* and
- given two disjoint intervals I_1, I_2,
$$E(\delta_{I_1}(\beta)\delta_{I_2}(\beta)) = 0. \tag{1.4}$$

It results from these two properties that β satisfies the dimension equation
$$\beta \simeq T^{\frac{1}{2}}. \tag{1.5}$$

2. Langevin Dynamic

We add to the Hamiltonian system a random force under which the equation (1.1) becomes
$$dp_i(t) + \frac{\partial V}{\partial q_i}dt = \sqrt{2\gamma_i\theta m_i}d\beta_i(t) - \gamma_i p_i(t)dt \qquad i \in [1, N]. \tag{2.1}$$

Here, the β_i are independent R^3-valued Brownian motions, γ_i is an experimental parameter called the *averaged oscillation frequency* of the i-th atom and $\theta = k^{-1}\times$ the temperature expressed in Kelvin (k denotes the constant of perfect gas appearing in the Boltzmann factor). The equation (2.1) implies the dimension equation $\gamma_i \simeq T^{-1}$. We call γ_i *the mean frequency of oscillation of the atom i.* As $\theta \simeq ML^2T^{-2}$, the required homogeneity appears as a whole in (2.1). The dynamic is obtained by solving together (2.1) and (1.2).

It will be proved (see Appendix A) that averaging during a long time along a single trajectory of the Langevin dynamic leads to the same result as that of averaging on the phase space for the Boltzmann measure $exp(-\frac{H}{\theta})dpdq$. This important fact justifies a-posteriori the introduction of a Langevin dynamic. Up to the choice of the averaged oscillation frequencies γ_i , there exists a unique Langevin dynamic so associated with the Boltzmann Statistic.

The Brownian motion is nowhere differentiable. For this reason equation (2.1) cannot be considered as an ODE (Ordinary Differential Equation): it is an SDE (Stochastic Differential Equation). This important fact can be left aside as a secondary technical question by using the *transfer principle* established in the reference P. Malliavin [2] Chapter VIII; this principle treats any SDE as a limit of highly oscillating ODEs.

3. Langevin Dynamic nearby the Equilibrium

We assume that at the point q_i^0 the potential reaches a local minimum. Then, there exists a quadratic form $\mathcal{Q}$ such that
$$V(q^0 + h) - V(q^0) = \mathcal{Q}(h) + o(h).$$

We denote the mass of the hydrogen atom by m_0; we introduce the quadratic form defined by

$$\mathcal{Q}'(m) = \sum_{i=1}^{N} \frac{m_i}{m_0} \|q_i\|^2_{R^3} .$$

Then, the quadratic form $\mathcal{Q}$ can be diagonalized in an orthonormal basis for $\mathcal{Q}'$. We call the sequence of positive numbers χ_i the *normal modes*, and $\frac{m_0}{2}\ \chi_i^2$ are the eigenvalues appearing in this diagonalization. The dimension equations are

$$V \simeq MLT^{-2}, \quad \mathcal{Q} \simeq ML^2T^{-2}, \quad \mathcal{Q}' \simeq L^2 \ ; \quad m_0\chi_i^2 L^2 \simeq ML^2T^{-2};$$

therefore $\chi_i \simeq T^{-1}$.

'Change of variables' simplifies the realization of diagonalization:

$$B_i = \sqrt{\frac{m_i}{m_0}}(q_i - q_i^0); \ then\ \mathcal{Q}' = \sum \|b_i\|^2_{R^3} .$$

Expressing $\mathcal{Q}$ in terms of the variables b_i reduces the problem at hand to a usual diagonalization problem in the new variables. We denote the matrix realizing the change of basis by α_*^*, and the new variables by $q_s^*,\ \ s \in [1, 3N]$. Then

$$q_s^* = \sum_{k=1}^{3N} \alpha_k^s b_k .$$

We consider the change of variables on the moments

$$a_i = \sqrt{\frac{m_0}{m_i}} p_i .$$

Then, the symplectic form is preserved by this change of variable; more precisely

$$\sum_i dp_i \wedge dq_i = \sum_i da_i \wedge db_i ,$$

which means that the change of variables $(p, q) \to (a, b)$ is a canonical transformation preserving the Hamiltonian formalism: define

$$H'(a, b) = \frac{1}{2m_0} \sum_{i=1}^{N} \|a_i\|^2_{R^3} + V(..., \sqrt{\frac{m_0}{m_i}} b_i, ...),$$

then, the Langevin dynamic can be written in the new coordinates as

$$da_i + \frac{\partial H'}{\partial b_i} dt = \sqrt{2\theta m_0}\sqrt{\gamma_i} d\beta_i - \gamma_i a_i dt; \qquad b_i = \frac{1}{m_0} a_i . \tag{3.1}$$

Define a change of variables on a as

$$p_s^* = \sum_{j=1}^{3N} \alpha_s^j a_j ; \qquad s \in [1, 3N], \tag{3.2}$$

then the symplectic form $\sum da \wedge db$ is preserved by this new change of variables. As before we deduce that in the new coordinates the Langevin dynamic is given by Hamiltonian formalism. It is easy to verify this fact by a brute force computation, and we get

Theorem 3.1. *Assume that all the averaged oscillation frequencies are equal; that is*

$$\gamma_i = \gamma, \qquad \forall i \in [1, N]. \tag{3.3}$$

Then there exist 3N independent Brownian motions such that the Langevin dynamic in the coordinates (p^*, q^*) is expressed by

$$dp_i^*(t) + m_0 \chi_i^2 q_i^* dt = \sqrt{2\gamma\theta m_0}\, d\beta_i^* - \gamma p_i^* dt; \qquad dq_i^* = \frac{1}{m_0} p_i^* dt \quad i \in [1, 3N]. \tag{3.4}$$

Proof. We define

$$\beta_i^* = \sum_{j=1}^{3N} \alpha_i^j \beta_j; \tag{3.5}$$

then the fact that the matrix α is orthogonal implies that the β_i^* are independent Brownian motions. □

Remark 3.2. *The motion of the protein in R^{6N} is decoupled into 3N independent motions, each motion having R^2 as phase space. Each of these motions will be integrated explicitly in the next section.*

4. Integration of the Langevin Motion of a Normal Mode

Here, for the mode of interest we omit the index i as the computation will be made for a particular i. In the phase space R^2, we consider the following Langevin dynamic:

$$dp = -(\gamma p + m_0 \chi^2 q)dt + \sigma d\beta, \; dq = pdt; \; (p, q) \in R^2, \; \text{where } \sigma = \sqrt{2\gamma\theta m_0}. \tag{4.1}$$

With the ordinary differential equation matrix in R^2,

$$dx = \mathcal{A}\, x\, dt, \qquad x = (p, q),$$
$$\mathcal{A} = \begin{bmatrix} -\gamma & -m_0\chi^2 \\ \frac{1}{m_0} & 0 \end{bmatrix},$$

and the first coordinate vector of R^2 denoted by e_1, the vector form of (4.1) can be given as :

$$dx_\beta = \mathcal{A} x_\beta \, dt + \sigma \, e_1 d\beta. \tag{4.2}$$

The characteristic equation of $\mathcal{A}$ is

$$\lambda^2 + \gamma\lambda + \chi^2 = 0,$$

and its discriminant is

$$\Delta = \gamma^2 - 4\chi^2.$$

Without any loss of generality, we can suppose that $\Delta < 0$; otherwise we shall be in a situation where the heat bath will induce stronger forces than the binding forces of the protein; this will yield immediate dissociation of the protein. The negative value of the discriminant implies that the eigenvalues of $\mathcal{A}$ are complex and have the following expression:

$$\begin{aligned} \lambda &= -\frac{\gamma}{2} + i\delta, \qquad \bar{\lambda} = -\frac{\gamma}{2} - i\delta, \qquad where, \\ \delta &= \sqrt{\chi^2 - \frac{\gamma^2}{4}}; \qquad therefore \qquad \delta \simeq T^{-1} \quad \lambda \simeq T^{-1}. \end{aligned}$$

We take

$$f = (1, \frac{1}{m_o \lambda}) \qquad and \qquad \bar{f} = (1, \frac{1}{m_o\, \bar{\lambda}})$$

as the corresponding eigenvectors. Since the real part of λ is $\mathcal{R}\lambda = -\frac{\gamma}{2} < 0$, we have

$$\|\exp(t\mathcal{A})\| \to 0 \ when \ t \to \infty. \tag{4.3}$$

The resolution of (4.2) is achieved by the Lagrange perturbative method: we introduce

$$y_\beta(t) = \exp(-t\mathcal{A})\ x_\beta(t).$$

Then, (4.2) can be written as

$$d\ y_\beta(t) = \sigma \exp(-t\mathcal{A})\ e_1 d\beta(t).$$

Therefore, the general solution of (4.2) is given by

$$x_\beta(t) = c \int_0^t \exp((t-s)\mathcal{A})\ e_1 d\beta(s) + \exp(t\mathcal{A})\ x_\beta(0). \tag{4.4}$$

The correlation is defined by the time average along a single trajectory of the Langevin dynamic

$$\mathcal{C}(\tau) = \lim_{R\to\infty} \int_0^R q_\beta(t)\ q_\beta(t+\tau)\ \frac{dt}{R}, \qquad \tau \geq 0. \tag{4.5}$$

Therefore, the determination of $\mathcal{C}(\tau)$ involves only the behaviour $x_\beta(t)$ for large time t; due to (4.3), the second term of (4.4) does not contribute to the computation of $\mathcal{C}(\tau)$. We shorten our notation by limiting ourselves to the case $x_\beta(0) = 0$.

Theorem 4.1. *Under the assumption that $x_\beta(0) = 0$, the formula (4.4) implies that*

$$q_\beta(t) = \Omega \int_0^t \Psi(t-s)\, d\beta(s) \tag{4.6}$$

where

$$\Psi_{\gamma,\delta}(v) = \exp(-\frac{\gamma}{2}v)\ \sin \delta v, \qquad \Omega = \frac{\sqrt{2\gamma\theta m_0}}{m_0\delta} \simeq LT^{\frac{-1}{2}}.$$

We introduce

$$Q(t) = \exp(\frac{\gamma}{2}t)\ \exp(t\mathcal{A}),$$

then $Q(t)$ has the following matrix form:

$$\begin{bmatrix} -\frac{\gamma}{2\delta}\sin\delta t + \cos\delta t & -\frac{m_0\chi^2}{\delta}\sin\delta t \\ \frac{1}{m_0\delta}\sin\delta t & \frac{\gamma}{2\delta}\sin\delta t + \cos\delta t \end{bmatrix}.$$

We let the matrix $\mathcal{S}$ have the column vectors f , $\bar{f}$, then

$$\mathcal{S} = \begin{bmatrix} 1 & 1 \\ \frac{1}{m_0\lambda} & \frac{1}{m_0\bar{\lambda}} \end{bmatrix}.$$

The matrix $\mathcal{S}^{-1}$ describes the change of basis : from the canonical basis of R^2 to the basis constituted by the two vectors f , $\bar{f}$. Direct computation gives

$$\mathcal{S}^{-1} = \begin{bmatrix} -i\frac{\lambda}{2\delta} & i\frac{\chi^2}{2\delta}m_0 \\ i\frac{\bar{\lambda}}{2\delta} & -i\frac{\chi^2}{2\delta}m_0 \end{bmatrix}.$$

We let $\mathcal{D}(t)$ denote the diagonal matrix

$$\mathcal{D}(t) = \begin{bmatrix} \exp(\lambda t) & 0 \\ 0 & \exp(\bar{\lambda}\, t) \end{bmatrix}.$$

Then,

$$\exp(tA) = SD(t)\, S^{-1}.$$

Denoting the vector transform of (p,q), $p \equiv p(t) q \equiv q(t)$, by $\exp(t\mathcal{A})$, letting

$$\varsigma = \frac{1}{2\theta}(-ip\lambda + i\chi^2 q)$$

as the second row of $\mathcal{S}^{-1}$ is the imaginary conjugate of the first one, we have

$$\mathcal{S}^{-1}(p,q) = (\varsigma, \bar{\varsigma});$$

therefore

$$\begin{aligned} p(t) &= 2\mathcal{R}(\exp(\lambda t)\varsigma) = \frac{\exp(-\frac{\gamma}{2}t)}{\delta}((-\frac{\gamma}{2}\sin\delta t + \delta\cos\delta t)p - (\chi^2\sin\delta t)q), \\ q(t) &= 2\mathcal{R}(\frac{1}{\lambda}\exp(\lambda t)\varsigma) = \frac{1}{\delta}\mathcal{R}(\exp(\lambda t)(-ip + i\,\bar{\lambda}\,q)) \\ &= \frac{\exp(-\frac{\gamma}{2}t)}{\delta}((\delta\cos\delta t + \frac{\gamma}{2}\sin\delta t)q + (\sin\delta t)p). \end{aligned}$$

Hence, we reread $Q(t)$. The proof of (4.6) is obtained by combining (4.4) with the expression of the term on the second row of the first column of the matrix $Q(t)$.

Theorem 4.2. *The correlation defined in (4.5) has its spectral density given by*

$$\Theta_{\gamma,\delta}(\omega) = \int_0^\infty C_{\gamma,\delta}(\tau)\cos(\tau\omega)d\tau = \frac{\theta}{m_0} \times \frac{\gamma}{(\omega^2-\chi^2)^2+\gamma^2\omega^2}. \tag{4.7}$$

Remark 4.3. *The dimension equation is* $\Theta_{\gamma,\delta} \simeq L^2 T$.

Proof. Define the Brownian motion β from $]-\infty\ to + \infty[$ and introduce the asymptotic process

$$q_\beta^\#(t) = \sigma\int_{-\infty}^{t} \Psi_{\gamma,\delta}(t-s)d\beta(s), \qquad t \in]-\infty\ ,+\infty[.$$

□

For $t>0$ and large, $q_\beta^\#(t)$ is a good approximation of $q_\beta(t)$. Furthermore, $q_\beta^\#$ is an ergodic gaussian process: time average is equal to average on the probability space. Therefore,

$$\begin{aligned} \Omega^{-2}C_{\gamma,\delta}(\tau) &= E(q^\#(0)q^\#(\tau)) \qquad (4.8) \\ &= \int_{-\infty}^{0} \Psi_{\gamma,\delta}(\tau-s)\Psi_{\gamma,\delta}(-s)ds, \end{aligned}$$

where the last equality is a consequence of (1.4).

Define

$$\begin{aligned} \tilde{\Psi}_{\gamma,\delta}(v) &= \{\Psi_{\gamma,\delta}(v)\ \ if\ \ v>0, \quad and\ 0\ \ if\ \ v\le 0\}, \\ \tilde{\Psi}^{\#}_{\gamma,\delta}(v) &= \tilde{\Psi}_{\gamma,\delta}(-v). \end{aligned}$$

Then, the correlation function can be expressed as the convolution product

$$\Omega^{-2}C_{\gamma,\delta}(\tau) = \tilde{\Psi}_{\gamma,\delta}(\tau) * \tilde{\Psi}^{\#}_{\gamma,\delta}(\tau). \qquad \tau>0.$$

The right-hand side of this equality is an even function; therefore, the cosine transform of the left-hand side is equal to $\frac{1}{2}$ the Fourier transform of the right-hand side; the Fourier transform of a convolution product is the product of the Fourier transform of each factor; the Fourier transform of $\tilde{\Psi}^{\#}_{\gamma,\delta}$ is the imaginary conjugate of the Fourier transform of $\tilde{\Psi}_{\gamma,\delta}$; therefore,

$$\begin{aligned}
\int_0^\infty \mathcal{C}_{\gamma,\delta}(\tau)\cos(\tau\omega)d\tau &= \frac{\Omega^2}{2}\left|\int_0^\infty \Psi(v)\exp(i\omega v)dv\right|^2, \\
\int_0^\infty \Psi(v)\exp(i\omega v)dv &= \int_0^\infty \exp(-\frac{\gamma}{2}v)\sin\delta v\exp(i\omega v)dv \\
&= \frac{1}{2i}\left[\frac{1}{\frac{\gamma}{2}-i\omega-i\delta}-\frac{1}{\frac{\gamma}{2}-i\omega+i\delta}\right] \\
&= \frac{\delta}{\left(\frac{\gamma}{2}-i\omega\right)^2+(\delta)^2}. \qquad (4.9)
\end{aligned}$$

Finally,

$$2\Theta_{\gamma,\delta}(\omega)=\Omega^2\left|\frac{\delta}{\left(\frac{\gamma}{2}-i\omega\right)^2+(\delta)^2}.\right|^2=\Omega^2\frac{\delta^2}{\left(\frac{\gamma^2}{4}-\omega^2+\delta^2\right)^2+\gamma^2\omega^2};$$

replacing δ^2 by $\chi^2-\frac{\gamma^2}{4}$, we get (4.7).

Higher order correlations: consider a continuous function of the two variables $\phi(\xi,\eta)$ and consider that the $\phi-$ *correlation* is given by

$$\begin{aligned}
\mathcal{C}_\phi(\tau) &= \lim_{R\to\infty}\int_0^R \phi(q^{\#}(t+\tau),q^{\#}(t))\frac{dt}{R} \\
&= \frac{1}{2\pi}\int_{R^2}\phi(\xi,\eta)exp(-\frac{1}{2}\mathcal{B}_{\gamma,\delta}(\xi,\eta))\frac{d\xi d\eta}{\sqrt{\det\mathcal{B}}}, \qquad (4.10)
\end{aligned}$$

where $\mathcal{B}^\tau_{\gamma,\delta}$ is the symmetric matrix which is the inverse of the symmetric matrix

$$\begin{bmatrix} \mathcal{C}_{\gamma,\delta}(0) & \mathcal{C}_{\gamma,\delta}(\tau) \\ \mathcal{C}_{\gamma,\delta}(\tau) & \mathcal{C}_{\gamma,\tau}(0) \end{bmatrix}.$$

When ϕ is a polynomial, it is classical that we can find a polynomial of two variables, P_ϕ, such that the required integral can be expressed as $P_\phi(\mathcal{C}_{\gamma,\delta}(0),\mathcal{C}_{\gamma,\delta}(\tau))$. The proof of (4.10) depends on the fact that the joint law of $q^{\#}(\tau),q^{\#}(0)$ is gaussian; such law is fully determined by its covariance matrix.

5. Computation of the spectral density of the motion of atoms

Let Γ be a given function on R^{6d}, and symmetric by the permutation of the group of variables of even rank with the group of variables of odd rank. We wish to compute the spectral measure associated with the statistics of $\Gamma(\ldots,q_i(\tau),q_j(0),\ldots)$. Then, by an affine change of vaiables, q can be expressed in terms of q^*; the function Γ expressed in the new variables is denoted by $\widetilde{\Gamma}$. Then, we have the exact value of the covariance

$$\int \widetilde{\Gamma}\,(....\xi_i,\eta_i,....)\prod_i \exp(-\frac{1}{2}\mathcal{B}^{\tau}_{\gamma,\delta_i}(\xi,\eta))\frac{d\xi_i d\eta_i}{2\pi\det(\mathcal{B}_{\gamma,\delta_i})}. \tag{5.1}$$

Basic knowledge needed to prove this formula is that the Brownians β^* are independent;this property is reflected in the fact that the gaussian measure is a product measure. We emphasize that this expression is exact. It is important to realize however that by replacing the potential V by its hessian, we have made a serious approximation. We may wonder whether it is so reasonable to compute (5.1) exactly; we can make an approximation by replacing $\widetilde{\Gamma}$ by its Taylor expansion up to the order $2r$. By reason of parity, all the monomials with an odd total degree will give no contribution to the integral. Taking $r = 1$, we get the following expression for the spectral density of the approximated covariance,

$$\frac{\theta}{m_0}\sum_i \frac{\partial^2\,\widetilde{\Gamma}}{\partial\xi_i\eta_i}(0)\frac{\delta_i^2}{(\chi_i^2-\omega^2)^2+\gamma^2\omega^2} \tag{5.2}$$

under the assumption that $\frac{\partial\widetilde{\Gamma}}{\partial\xi_i^2}(0) = 0$.

This formula superimposes functions which have peaks at principal modes. If we take $r = 2$, then the fourth derivatives of $\widetilde{\Gamma}$ have to appear; among their coefficients, we find expressions of the form $\Theta_{\gamma\delta_i} * \Theta_{\gamma\delta_j}$ which have the inconvenience to produce a new peak at the point $\chi_i + \chi_j$.

6. Computation of the correlations under discrepancies in the mobility

As is described in the Appendix, under the Langevin paradigm and when we know the temperature, there exists only one undetermined constant for each atom: a constant that we denote by γ_i. This has the dimension $\gamma_i \simeq T^{-1}$. We resume the computation by assuming that γ_i are distinct. We start from equation (3.1), where each Brownian is multiplied by $\sqrt{\sigma}\sqrt{\gamma_i}$, $\sigma = 2\theta m_0$. We denote the matrix of change of orthonormal basis by α. Then,

$$\beta_s^* = \sqrt{\sigma}\sum_k \alpha_s^k\sqrt{\gamma_k}\beta_k.$$

The β_s^* are Brownian motions up to a constant factor; but they are not independent anymore. As the Brownian motions are gaussian processes, all the structure is determined by the covariance

$$\sigma^{-1} < d\,\beta_s^*, d\beta_q^* >= \sum_k \alpha_s^k\alpha_q^k\gamma_k = \kappa_{s,q}; \qquad \kappa_{s,q} \simeq T^{-1}.$$

When all the γ_k are equal, we find that this sum is the Kronecker delta δ_s^q. The spectral functions of q_s^* are still given by (4.7) multiplied by $\kappa_{q,q}$. Formula (5.1) is no longer true.

Let us compute the spectral density Θ of the covariance of a linear form

$$\Phi(q) = \sum_i c_i q_i^*,$$

c_i : constants.

Denote $\widetilde{\kappa} = \sqrt{\kappa}$, with the square root taken in the sense of matrix calculus. Then we have the following theorem:

Theorem 6.1.

$$\Theta(\omega) = \frac{\theta}{m_0} \times \sum_r \left| \sum_s \widetilde{\kappa}_{r,s}\, c_s \frac{\delta_s}{(\frac{\gamma_s}{2} - i\omega)^2 + \delta_s^2} \right|^2,$$

where $\delta_s = \sqrt{\chi_s^2 - \frac{\gamma_s^2}{4}}$.

The proof of this theorem is based on formula (4.8).

7. Appendix A: Langevin dynamic and Boltzmann Statistic

We associate with the Hamiltonian

$$H(p,q) = \sum_i \frac{1}{2m_i} \|p_i\|_{R^3}^2 + V(q) \qquad where \qquad (p,q) \in R^{6d}$$

the Langevin equation

$$dp_i = -\frac{\partial H}{\partial q}\, dt + c_i\, d\beta_i - \gamma_i p\, dt, \qquad dq = \frac{\partial H}{\partial p}\, dt,$$

where β_i is an R^{3d}-valued Brownian motion and where the constants c_i , γ_i will be determined later.

Given a C^2 function Φ defined on the phase space, the infinitesimal generator associated with the Langevin dynamic is defined by

$$\lim_{R\to\infty} \epsilon^{-1} E(\Phi(p(\epsilon), q(\epsilon)) - \Phi(p(0), q(0))) = \chi\Phi = \mathcal{L}\Phi + \{H, \Phi\} \tag{7.1}$$

where

$$\{H, \Phi\} = \sum_{k=1}^{d} \left(\frac{\partial H}{\partial q_k}\frac{\partial \Phi}{\partial p_k} - \frac{\partial H}{\partial p_k}\frac{\partial \Phi}{\partial q_k}\right)$$

is the Poisson Bracket, and

$$\mathcal{L}\Phi = \sum_{i=1}^{d} \mathcal{L}_i \;\; with \qquad \mathcal{L}_i = \frac{c_i^2}{2}\Delta_i - \gamma_i (p_i \mid \nabla\Phi), \qquad \Delta_i = \sum_{k=1}^{3} \frac{\partial^2}{\partial (p_i^k)^2}.$$

The Poisson Bracket is a *derivation*: namely, it satisfies the identity

$$\{H, (\Phi\Psi)\} = \Psi\{H, \Phi\} + \Phi\{H, \Psi\}. \tag{7.2}$$

Furthermore, the Liouville measure $dp \wedge dq$ is preserved under the Hamiltonian flow which means that

$$\int_{R^{3d}} \int_{R^{3d}} \{H, \Phi\} \prod_i dp_i \prod_i dq_i = 0. \tag{7.3}$$

Theorem 7.1. *The Boltzmann measure*

$$\nu = \exp(-\frac{1}{\theta} H) \prod_i dp_i \prod_i dq_i$$

is invariant under the Langevin dynamic if and only if (7.2) is fulfilled.

Proof. The invariance is equivalent to the fact that for every C^2 function Φ we have

$$\int_{R^{6d}} \chi \Phi dv = 0. \tag{7.4}$$

We next show that the two terms of the decomposition (7.1) satisfy (7.4): according to (7.3)

$$\int_{R^{6d}} \{H, \exp(-\beta H) \Phi\} dp \wedge dq = 0.$$

As, by (7.2) $\{H, \exp(-\frac{1}{\theta}H)\} = -\beta \exp(-\frac{1}{\theta}H)\{H, H\} = 0$, we get

$$\int_{R^{6d}} \{H, \Phi\} \exp(-\frac{1}{\theta} H) dp \wedge dq = 0. \tag{7.5}$$

We remark that $\mathcal{L}$ contains derivatives only in the variables p; we shall integrate firstly with respect to p :

$$\int_{R^{6d}} \mathcal{L}\Phi \exp(-\frac{1}{\theta}H) dp \wedge dq = \int_{R^{3d}} \exp(-\frac{1}{\theta}V) dq \Big(\int_{R^{3d}} \exp\Big(-\frac{1}{\theta} \sum_k \frac{\|p_k\|^2}{m_k}\Big) \mathcal{L}\Phi dp \Big).$$

The term in parentheses is treated by splitting $\mathcal{L} = \sum_i \mathcal{L}_i$; we are now to do integration by parts on the variable $p_i \in R^3$, in order to show that

$$\int_{R^3} (\frac{c_i^2}{2} \Delta_i \phi - \gamma_i (\xi \mid \nabla_i \phi)_{R^3}) \exp(-\frac{1}{2\theta m_i} \|\xi\|^2_{R^3}) d\xi = 0. \tag{7.6}$$

By splitting on each of the three coordinates of R^3, the integration to be performed reduces to integration by parts on R; we introduce the following *symmetric* bilinear form:

$$\frac{1}{2} \int_R c_i^2 y' v' \exp(-\frac{\xi^2}{2m_i \theta}) d\xi = \int_R (-\frac{1}{2} c_i^2 y'' + \frac{1}{2} c_i^2 \frac{1}{\theta m_i} \xi y') v \exp(-\frac{1}{2m_i \theta} \xi^2) d\xi.$$

Taking the identity $v = 1$ and if also the relation

$$c_i^2 = 2\gamma_i \theta m_i \tag{7.7}$$

is satisfied, we obtain (7.6). □

References

[1] King R. and Jardetzky O. *A general formalism for the Analysis of NMR relaxation measurements on system with multiple degree of Freedom.* Chemical Physics Letters **53** (1978) 15–18.

[2] Malliavin P. *Stochastic Analysis.* Springer Verlag. 1997.

[3] Ribeiro A., King R., Restivo C. and Jardetzky O. *An approach to the mapping of internal motions in Proteins. Analysis of* ^{13}C *NMR Relaxation in the Bovine Panereatic Trypsin Inhibitor.* J. American Chemical Society. 162–12 (1980) 4040–4051.

10, Rue Saint Louis en L'Isle
75004 Paris
France

Department of Mathematics
Eastern Mediterranean University
G. Magusa
North Cyprus (via Mersin 10 – Turkey)

Progress in Probability, Vol. 53, 181–202

Quasi-invariance for Lévy Processes under Anticipating Shifts

Nicolas Privault

Abstract. We prove a Girsanov theorem for the combination of a Brownian motion on R_+ and a Poisson random measure on $R_+ \times [-1,1]^d$ under random anticipating transformations of paths and configurations. The factorization of the density function via Carleman–Fredholm determinants and divergence operators appears as an extension of the martingale factorization in the adapted jump case.

1. Introduction

The Cameron–Martin theorem [5] gives the density with respect to the Wiener measure of a deterministic shift of Brownian motion. Similarly, the Skorokhod theorem on invariance of measures [18] gives the density with respect to Poisson measures of deterministic shifts of configuration. These theorems have an extension (the Girsanov theorem) to random shifts under adaptedness hypothesis. Given a martingale $(\tilde{M}(t))_{t\in\mathbb{R}_+}$ on a filtered probability space $(\Omega, \mathcal{F}, (\mathcal{F}_t)_{t\in\mathbb{R}_+}, P)$ and a probability Q absolutely continuous with respect to P, the classical Girsanov theorem, [7], [8], gives a canonical decomposition of $(\tilde{M}(t))_{t\in\mathbb{R}_+}$ as a sum of a continuous martingale $(M^c(t))_{t\in\mathbb{R}_+}$ under Q, a pure jump martingale $(M^d(t))_{t\in\mathbb{R}_+}$ under Q and a finite variation process. Since $(M^c(t))_{t\in\mathbb{R}_+}$ and $(M^d(t))_{t\in\mathbb{R}_+}$ are obtained by shifts of the values of $(\tilde{M}(t))_{t\in\mathbb{R}_+}$, the Girsanov theorem also allows us to compute the density with respect to P of a transformation of the space Ω. Thus there is a strong analogy between the Girsanov theorem and change of variable formulas in the theory of integration, the density of P with respect to Q being computed with a Jacobian determinant. The classical Girsanov theorem relies on Itô's stochastic calculus, and in particular on adaptedness hypotheses which are not needed in integration (change of variable formula) techniques. The latter point of view has proved to be useful to remove the adaptedness conditions imposed by the Girsanov theorem on transformations of trajectories. In order to deal with stochastic processes, a theory of integration in infinite dimensions is needed, and analysis on the Wiener space provides such a framework in the case of Brownian motion. The extension of the Girsanov theorem to anticipating shifts of Brownian motion has been carried out in [10], [17], [19], [20], see the book

[21] for more complete references. In the standard Poisson case (i.e., for Poisson random measures based on $\mathbb{R}_+$), its analog has been treated in [14], [16], using analysis for an infinite product of exponential densities. This result relies on the interpretation of Poisson samples on $\mathbb{R}_+$ as sequences of independent exponentially distributed interjump distances. On the other hand, a change of variable formula for the uniform density in infinite dimensions has been established in [15].

In this paper we obtain a Girsanov type theorem for random shifts of a Poisson random measure on $\mathbb{R}_+ \times [-1,1]^d$ and a Brownian motion, i.e., an anticipative Girsanov theorem for Lévy processes. The main observation is that a Poisson random measure on $\mathbb{R}_+ \times [-1,1]^d$ with flat intensity consists in randomly distributed sets of points (configurations) that can be represented as sequences of independent $d+1$-dimensional random variables whose first component is exponentially distributed, the remaining independent d components having uniform laws on $[-1,1]$. The Radon–Nikodym density function is then factorized with a divergence operator and a Carleman–Fredholm determinant and we allow for interactions between the different components of the process. This factorization of the density is similar to the expression of the density via stochastic calculus, as the solution of a stochastic differential equation. Girsanov type theorems for non-adapted shifts of Poisson random measures are completely natural since in the Poisson case on $\mathbb{R}^{d+1}$ there is no canonical notion of time or filtration.

We proceed as follows. In Sect. 2 we review different versions of the adapted Girsanov theorem. In Sect. 3 we introduce some notation, in particular an interpretation of the Poisson space as a space of sequences. The main results are stated in Sect. 4, for shifts of configuration points that are expressed as perturbations of interjump times and jump heights. In Sect. 5 and Sect. 6 we prove technical results and then our extension of the Girsanov theorem to the anticipating case.

2. Adapted Girsanov theorem and change of variable formulas

In this section we review the classical Skorokhod theorem on the absolute continuity of Poisson measures under deterministic diffeomorphisms, cf. [1], [18], [22], and its extension to adapted shifts, i.e., the Girsanov theorem, cf. [7], [8]. In Sect. 4 we will present an extension of this theorem to the anticipating case.

Let $\Gamma(\mathbb{R}_+ \times [-1,1]^d)$ denote the configuration space on $\mathbb{R}_+ \times [-1,1]^d$, i.e., the set of Radon measures on $\mathbb{R}_+ \times [-1,1]^d$ of the form

$$\sum_{k=1}^{k=N} \varepsilon_{x_k}, \quad (x_k)_{k=1}^{k=N} \subset \mathbb{R}_+ \times [-1,1]^d, \; x_k \neq x_l, \; \forall k \neq l, \; N \in \mathbb{N} \cup \{\infty\}.$$

A configuration γ is a sum of Dirac measures ε_y and will be identified with the discrete set of points that defines its support, in particular we will write $(s,y) \in \gamma$ whenever $\gamma(\{(s,y)\}) = 1$. Let $\mathcal{C}_0(\mathbb{R}_+)$ denote the space of continuous functions starting at 0.

Let $\Omega = \mathcal{C}_0(\mathbb{R}_+) \times \Gamma(\mathbb{R}_+ \times [-1,1]^d)$ and consider a filtered probability space $(\Omega, \mathcal{F}, (\mathcal{F}_t)_{t\in\mathbb{R}_+}, P)$.

Let ν be a diffuse random measure on $\mathbb{R}_+ \times [-1,1]^d$, bounded on compact sets, $\mathcal{F}_t$-predictable under P, i.e., $t \mapsto \int_0^t \int_{[-1,1]^d} u(s,y)\nu(ds,dy)$ is $\mathcal{F}_t$-predictable for every positive bounded $\mathcal{F}_t$-predictable process $(u(s,y))_{(s,y)\in\mathbb{R}_+\times[-1,1]^d}$. We assume that the canonical continuous process $t \mapsto \omega(t)$ defined on $\mathcal{C}_0(\mathbb{R}_+)$ has an $\mathcal{F}_t$-predictable quadratic variation $(\beta(t))_{t\in\mathbb{R}_+}$ under P, and that ν is the intensity (or Lévy measure, or dual predictable projection) of the random measure $\mu : \Omega \longrightarrow \Gamma(\mathbb{R}_+ \times [-1,1]^d)$ defined as $\mu(\omega,\gamma) = \gamma$, i.e., $\nu(ds,dy)$ satisfies

$$E_P\left[\int_{\mathbb{R}_+\times[-1,1]^d} u(s,y)\gamma(ds,dy)\right] = E_P\left[\int_{\mathbb{R}_+\times[-1,1]^d} u(s,y)\nu(ds,dy)\right],$$

for every positive bounded $\mathcal{F}_t$-predictable process $(u(s,y))_{(s,y)\in\mathbb{R}_+\times[-1,1]^d}$. Let $\phi : \mathbb{R}_+ \times [-1,1]^d \longrightarrow \mathbb{R}_+ \times [-1,1]^d$ be a random diffeomorphism of $\mathbb{R}_+ \times [-1,1]^d$ whose $d+1$ components

$$(\phi^1(t,y))_{(t,y)\in\mathbb{R}_+\times[-1,1]^d}, \ldots, (\phi^{d+1}(t,y))_{(t,y)\in\mathbb{R}_+\times[-1,1]^d},$$

are $\mathcal{F}_t$-predictable processes. We assume that there exists a predictable process $(Z(s,y))_{(s,y)\in\mathbb{R}_+\times[-1,1]^d} > -1$, P a.s., such that

(i) $Z \circ \phi \in L^1(\mathbb{R}_+ \times [-1,1]^d, \nu)$, i.e.,

$$\int_{\mathbb{R}_+\times[-1,1]^d} |Z \circ \phi(s,y))|\nu(ds,dy) < \infty, \quad P - a.s,$$

(ii) and

$$\int_{\mathbb{R}_+\times[-1,1]^d} g(\phi(s,y))(1+Z\circ\phi(s,y))\nu(ds,dy) = \int_{\mathbb{R}_+\times[-1,1]^d} g(s,y)\nu(ds,dy), \quad (2.1)$$

$\forall g \in \mathcal{C}_c^+(\mathbb{R}^{d+2})$, i.e., $(1+Z)^{-1}$ is the density of $\phi_*\nu$ with respect to ν, P-a.s.

If ν is a multiple of the Lebesgue measure, then $1 + Z \circ \phi(s,y) = |\partial\phi(s,y)|$ is the Jacobian determinant of ϕ. Let $\phi(\gamma)$ denote the configuration γ whose points are shifted according to ϕ, i.e., $\phi(\gamma)$ is identified with the set

$$\phi(\gamma) = \{\phi((s,y)) \ : \ (s,y) \in \gamma\}, \quad \gamma \in \Gamma(\mathbb{R}_+ \times [-1,1]^d).$$

Let $(z(s))_{s\in\mathbb{R}_+} \in L^2(\Omega) \otimes L^2(\mathbb{R}_+, d\beta)$ be a square-integrable $\mathcal{F}_t$-predictable process and let $\psi(\omega)$ be defined as

$$\psi(\omega)(t) = \omega(t) + \int_0^t z(s)d\beta(s), \quad t \in \mathbb{R}_+, \ \omega \in \mathcal{C}_0(\mathbb{R}_+).$$

In the following result, $(M^c(t))_{t\in\mathbb{R}_+}$ is a time changed Brownian motion and μ is a Poisson random measure with deterministic intensity ν.

Theorem 2.1. *Assume that ν and β are deterministic, and*

$$E_P\left[\exp\left(-\int_0^\infty z(s)d\omega(s) - \int_{\mathbb{R}^d} Z\circ\phi(s,y)\nu(ds,dy) - \frac{1}{2}\int_0^\infty z^2(s)d\beta(s)\right)\right.$$
$$\left.\times\prod_{(s,y)\in\gamma}(1+Z\circ\phi(s,y))\right] = 1. \tag{2.2}$$

Then for every bounded and measurable random variable $f:\Omega\longrightarrow\mathbb{R}$ we have

$$E_P[f(\omega,\gamma)] = E_P\left[f(\psi(\omega),\phi(\gamma))\exp\left(-\int_0^\infty z(s)d\omega(s) - \int_{\mathbb{R}^d} Z\circ\phi(s,y)\nu(ds,dy)\right.\right.$$
$$\left.\left.-\frac{1}{2}\int_0^\infty z^2(s)d\beta(s)\right)\prod_{(s,y)\in\gamma}(1+Z\circ\phi(s,y))\right].$$

We can also write

$$\frac{d\Phi_*^{-1}P}{dP} = \exp\left(-\int_0^\infty z(s)d\omega(s) + \int_{\mathbb{R}^d} Z\circ\phi(s,y)\nu(ds,dy) - \frac{1}{2}\int_0^t z^2(s)d\beta(s)\right)$$
$$\times\prod_{(s,y)\in\gamma}(1+Z\circ\phi(s,y)),$$

where the transformation $\Phi:\Omega\longrightarrow\Omega$ is defined as $\Phi(\omega,\gamma)=(\psi(\omega),\phi(\gamma))$, $(\omega,\gamma)\in\Omega$. We will recall a proof of the above result using an extension of the classical Girsanov theorem to the jump case in the martingale framework, cf. [8], [9]:

Theorem 2.2. *Let $(\tilde{M}(t))_{t\in\mathbb{R}_+}$ be a martingale on $(\Omega,\mathcal{F},P)$, whose continuous part $(\tilde{M}^c(t))_{t\in\mathbb{R}_+}$ has quadratic variation $(\tilde{\beta}(t))_{t\in\mathbb{R}_+}$. Assume that the jump part $(\tilde{M}^d(t))_{t\in\mathbb{R}_+}$ of $(\tilde{M}(t))_{t\in\mathbb{R}_+}$ is given by the discrete random measure $\tilde{\mu}:\Omega\longrightarrow\Gamma(\mathbb{R}_+\times[-1,1]^d)$ with intensity $\tilde{\nu}$ as*

$$\tilde{M}^d(t) = \int_0^t\int_{[-1,1]^d}\tilde{\mu}(ds,dy) - \int_0^t\int_{[-1,1]^d}\tilde{\nu}(ds,dy), \quad t\in\mathbb{R}_+.$$

Let

$$(\tilde{Z}(s,y))_{(s,y)\in\mathbb{R}_+\times[-1,1]^d}\in L^1(\mathbb{R}_+\times[-1,1]^d,\tilde{\nu}), \quad \textit{with } \tilde{Z}>-1, \quad P-a.s.,$$

and $(\tilde{z}(s))_{s\in\mathbb{R}_+}\in L^2(\Omega)\otimes L^2(\mathbb{R}_+,d\tilde{\beta})$ be predictable processes, and define the measure Q by its density

$$\frac{dQ}{dP} = \exp\left(-\int_0^\infty\tilde{z}(s)d\tilde{M}^c(s) - \int_{\mathbb{R}_+\times[-1,1]^d}\tilde{Z}(s,y)\tilde{\nu}(ds,dy) - \frac{1}{2}\int_0^\infty\tilde{z}^2(s)d\tilde{\beta}(s)\right)$$
$$\times\prod_{(s,y)}(1+\tilde{Z}(s,y)\Delta\tilde{M}(s,y)), \tag{2.3}$$

where $\Delta\tilde{M}(s,y)=\tilde{\mu}(\{(s,y)\})$, $(s,y)\in\mathbb{R}_+\times[-1,1]^d$.

If Q is a probability measure, then, under Q, the process $(\tilde{M}(t))_{t\in\mathbb{R}_+}$ admits a unique decomposition

$$\tilde{M}(t) = M^c(t) + M^d(t) + \alpha(t), \quad t \in \mathbb{R}_+,$$

where

(i) $M^c(t) = \tilde{M}^c(t) + \int_0^t \tilde{z}(s)d\beta(s)$, $t \in \mathbb{R}_+$, *is a continuous local martingale under Q with predictable quadratic variation $\beta^c(t) = \tilde{\beta}^c(t)$, $t \in \mathbb{R}_+$,*

(ii) $M^d(t) = \tilde{M}^d(t) - \int_0^t \int_{[-1,1]^d} \tilde{Z}(s,y)\tilde{\nu}(ds,dy)$, $t \in \mathbb{R}_+$, *is a pure jump martingale and $d\tilde{\mu}$ has intensity $(1+\tilde{Z})d\tilde{\nu}$ under Q,*

(iii) $\alpha(t) = \int_0^t \int_{[-1,1]^d} \tilde{Z}(s,y)\tilde{\nu}(ds,dy) - \int_0^t \tilde{z}(s)d\beta(s)$, $t \in \mathbb{R}_+$, *is a finite variation process.*

Proof. of Th. 2.1 from Th. 2.2. let $\tilde{Z} = Z$, $\tilde{z} = z$, assume that $(\tilde{M}(t))_{t\in\mathbb{R}_+} = (\tilde{M}^d(t))_{t\in\mathbb{R}_+} + (\tilde{M}^c(t))_{t\in\mathbb{R}_+}$ is defined as $\tilde{M}^c(t) = \omega(t)$ and

$$\tilde{M}^d(t) = \int_0^t \int_{[-1,1]^d} \phi(\gamma)(ds,dy) - \int_0^t \int_{[-1,1]^d} \tilde{\nu}(ds,dy), \quad t \in \mathbb{R}_+,$$

with $\tilde{\mu}(\omega,\gamma) = \phi(\gamma)$ and $\tilde{\nu}(ds,dy) = (1+Z(s,y))^{-1}\nu(ds,dy)$. Let $(\tilde{\mathcal{F}}_t)_{t\in\mathbb{R}_+}$ denote the smallest filtration that makes every process $(u(s,y))_{(s,y)\in\mathbb{R}_+\times[-1,1]^d}$ adapted if $(u\circ\phi(s,y))_{(s,y)\in\mathbb{R}_+\times[-1,1]^d}$ is $\mathcal{F}_t$-adapted. The process $(\tilde{M}(t))_{t\in\mathbb{R}_+}$ is $(\tilde{\mathcal{F}}_t)_{t\in\mathbb{R}_+}$ adapted. Given a positive bounded $\tilde{\mathcal{F}}_t$-predictable process, $(u(s,y))_{(s,y)\in\mathbb{R}_+\times[-1,1]^d}$, we have

$$\begin{aligned}
E_P\left[\int_{\mathbb{R}_+\times[-1,1]^d} u(s,y)\ \phi(\gamma)(ds,dy)\right] &= E_P\left[\int_{\mathbb{R}_+\times[-1,1]^d} u\circ\phi(s,y)\ \gamma(ds,dy)\right] \\
&= E_P\left[\int_{\mathbb{R}_+\times[-1,1]^d} u\circ\phi(s,y)\ \nu(ds,dy)\right] \\
&= E_P\left[\int_{\mathbb{R}_+\times[-1,1]^d} \frac{u(s,y)}{1+Z(s,y)}\ \nu(ds,dy)\right] \\
&= E_P\left[\int_{\mathbb{R}_+\times[-1,1]^d} u(s,y)\ \tilde{\nu}(ds,dy)\right].
\end{aligned}$$

Hence under P, $\tilde{\mu}$ has intensity $\tilde{\nu}$, and $(\tilde{M}^c(t))_{t\in\mathbb{R}_+}$ has quadratic variation $(\tilde{\beta}(t))_{t\in\mathbb{R}_+} = (\beta(t))_{t\in\mathbb{R}_+}$. Let $(M^c(t))_{t\in\mathbb{R}_+}$ be defined as $M^c(t) = \psi(\omega)(t)$, $t \in \mathbb{R}_+$, and let

$$M^d(t) = \int_0^t \int_{[-1,1]^d} \tilde{\mu}(ds,dy) - \int_0^t \int_{[-1,1]^d} \nu(ds,dy).$$

From Th. 2.2, $(M(t))_{t\in\mathbb{R}_+}$ is a martingale under Q, the random measure $\tilde{\mu}$ has intensity ν under Q, and $\psi(\omega)$ is a continuous martingale under Q with quadratic variation $\tilde{\beta}$. Hence $\phi(\gamma)$ and $\psi(\omega)$ have the same laws under Q as γ and ω under

P, since β and ν are deterministic. It remains to notice that (2.3) can be rewritten as

$$\frac{dQ}{dP} = \exp\left(- \int_0^{\infty} z(s) d\omega(s) - \int_{\mathbb{R}_+ \times [-1,1]^d} Z \circ \phi(s,y) \nu(ds,dy) - \frac{1}{2} \int_0^t z^2(s) d\beta(s) \right) \times \prod_{(s,y)\in\gamma} (1 + Z \circ \phi(s,y)), \tag{2.4}$$

since

$$\Delta\tilde{M}(\tilde{s},\tilde{y}) = \begin{cases} 1 & \text{if } \exists\ (s,y) \in \gamma\ :\ (\tilde{s},\tilde{y}) = \phi(s,y), \\ 0 & \text{otherwise,} \end{cases}$$

and

$$\begin{aligned} \int_{\mathbb{R}_+ \times [-1,1]^d} Z(s,y)\tilde{\nu}(ds,dy) &= \int_{\mathbb{R}_+ \times [-1,1]^d} \frac{Z(s,y)}{1 + Z(s,y)} \nu(ds,dy) \\ &= \int_{\mathbb{R}_+ \times [-1,1]^d} Z \circ \phi(s,y) \nu(ds,dy). \end{aligned}$$

□

We make some remarks on different factorizations of the density function and on the deterministic case.

- If the process Z does not belong to $L^1(\mathbb{R}_+ \times [-1,1]^d, \tilde{\nu})$, P-a.s., then the product (2.4) may not converge. However, if Z is predictable and belongs to $L^2(\Omega) \otimes L^2(\mathbb{R}_+ \times [-1,1]^d, \tilde{\nu})$, i.e.,

$$E_P\left[\int_{\mathbb{R}_+ \times [-1,1]^d} |Z \circ \phi(s,y)|^2 \nu(ds,dy) \right] = E_P\left[\int_{\mathbb{R}_+ \times [-1,1]^d} |Z(s,y)|^2 \tilde{\nu}(ds,dy) \right] < \infty,$$

then the probability density dQ/dP can still be formulated via a "renormalization" due to stochastic calculus. Namely, it can be written as the limit L_∞ as t goes to infinity of the process

$$L_t = \exp\left(- \int_0^t z(s) d\tilde{M}^c(s) + \int_0^t \int_{[-1,1]^d} Z(s,y)(\tilde{\mu}(ds,dy) - \tilde{\nu}(ds,dy)) - \frac{1}{2} \int_0^t z^2(s) d\tilde{\beta}(s) \right) \prod_{(s,y),\ s \le t} (1 + Z(s,y)\Delta\tilde{M}(s,y)) e^{-Z(s,y)\Delta\tilde{M}(s,y)}, \tag{2.5}$$

$t \in \mathbb{R}_+$, which is a solution to the Itô stochastic differential equation

$$\frac{1}{L_{t^-}} dL_t = z(t) d\tilde{M}^c(t) + \int_{[-1,1]^d} Z(t,y)(\tilde{\mu}(dt,dy) - \nu(dt,dy)), \quad L_0 = 1, \quad t > 0.$$

The factorization

$$\prod_{(s,y),\ s \le t} (1 + Z(s,y)\Delta\tilde{M}(s,y)) e^{-Z(s,y)\Delta\tilde{M}(s,y)} \tag{2.6}$$

is used in (2.5) because this modified product still converges as $t \longrightarrow \infty$ when $Z \in L^2(\Omega) \otimes L^2(\mathbb{R}_+ \times [-1,1]^d, \tilde{\nu})$ since $P(d\gamma)$ almost surely, the series

$$\sum_{(s,y)\in\gamma} (Z \circ \phi(s,y))^2$$

is summable. Moreover, in (2.5) the stochastic integral

$$\int_0^t \int_{[-1,1]^d} Z(s,y)(\tilde{\mu}(ds,dy) - \tilde{\nu}(ds,dy))$$

makes sense in $L^2(\Omega)$ for predictable $Z \in L^2(\Omega) \otimes L^2(\mathbb{R}_+ \times [-1,1]^d, \tilde{\nu})$. The same factorization will be used in the anticipating case, where the traditional determinant is replaced by a Carleman–Fredholm determinant (see (4.2)), compensated with a divergence operator, cf. [17], [23] (see below in Sect. 4).

- The absolute continuity result for deterministic shifts on Poisson space of [3], [18], [22] follows from the Girsanov theorem, in the particular case of a smooth deterministic diffeomorphiom

$$\phi : \mathbb{R}_+ \times [-1,1]^d \longrightarrow \mathbb{R}_+ \times [-1,1]^d.$$

We denote by P_ν the Poisson measure on $\Gamma(\mathbb{R}_+ \times [-1,1]^d)$ with deterministic intensity ν, i.e., the probability under which $\mu(\omega,\gamma) = \gamma$ has intensity ν, with characteristic function

$$E_{P_\nu}\left[\exp\left(\int_{\mathbb{R}_+\times[-1,1]^d} g d\gamma\right)\right] = \exp\left(\int_{\mathbb{R}_+\times[-1,1]^d} (e^{g(y)} - 1)\nu(dy)\right),$$

$g \in \mathcal{C}_c(\mathbb{R}_+ \times [-1,1]^d)$. Th. 2.1 gives

$$E_P[f(\omega,\gamma)] = E_P\left[f(\omega,\phi(\gamma))\exp\left(-\int_{\mathbb{R}^d} Z\circ\phi(s,y)\nu(ds,dy)\right)\prod_{(s,y)\in\gamma}(1 + Z\circ\phi(s,y))\right],$$

$f : \Omega \longrightarrow \mathbb{R}$ measurable bounded, where $(1+Z)^{-1}$ is the density of $\phi_*\nu$ with respect to ν. In the particular case where ν is the Lebesgue measure on $\mathbb{R}_+ \times [-1,1]^d$ we find:

$$\frac{d\phi_*^{-1}P_\nu}{dP_\nu} = \exp\left(\int_{\mathbb{R}_+\times[-1,1]^d} (|\partial\phi(y)| - 1)dy\right)\prod_{(s,y)\in\gamma} |\partial\phi(s,y)|.$$

3. Sequence model for the space Ω

In this section we introduce the sequence model of Ω as a vector space denoted by B, i.e., the random element (ω,γ) is constructed via a sequence of vectors which have independent Gaussian, exponential and uniformly distributed components.

In the remainder of this paper we work in the case of a deterministic flat intensity ν given as

$$d\nu(y^1,\dots,y^{d+1}) = \frac{1}{2^d} 1_{\mathbb{R}_+}(y^1) 1_{[-1,1]}(y^2)\cdots 1_{[-1,1]}(y^{d+1}) dy^1 \cdots dy^{d+1}, \tag{3.1}$$

and with the quadratic variation $\beta(t) = t$, $t \in \mathbb{R}_+$. (We will use the notation $\mathbb{R}^{d+2} \ni x_k = (x_k^0,\dots,x_k^{d+1})$). Let $B = \{x = (x_k)_{k\in\mathbb{N}} : x_k \in \mathbb{R}^{d+2}\}$, with the norm

$$\|x - y\|_B = \sup_{k\geq 0} \|x_k - y_k\|_{\mathbb{R}^{d+2}}/(k+1),$$

and associated Borel σ-algebra $\mathcal{F}$. Let $d \geq 1$, $r \geq 0$, and let λ_r be the finite measure on $\mathbb{R}^{d+2}$ defined by its density

$$d\lambda_r(t_0,t_1,\dots,t_{d+1} = \frac{1}{2^{d-1}\sqrt{2\pi}} e^{-t_0^2/2} e^{-t_1} 1_{\mathbb{R}}(t_0) 1_{[-r,\infty[}(t_1) 1_{[-r-1,1+r]}(t_2)\cdots$$
$$\cdots 1_{[-r-1,1+r]}(t_{d+1}) dt_0 \cdots dt_{d+1}.$$

We denote by P the probability defined on $(B,\mathcal{F})$ via its expression on cylinder sets:

$$P(\{x = (x_k)_{k\in\mathbb{N}} \in B \ : \ (x_0,\dots,x_n) \in A\}) = \lambda_0^{\otimes n+1}(A), \tag{3.2}$$

where A is a Borel set in $(\mathbb{R}^{d+2})^{n+1}$, $n \in \mathbb{N}$. We denote by

$$\tau_k = (\tau_k^0,\dots,\tau_k^{d+1}) : B \longrightarrow \mathbb{R}^{d+2} \quad k \in \mathbb{N},$$

the coordinate functionals defined as

$$\tau_k(x) = x_k = (x_k^0,\dots,x_k^{d+1}),$$

and

$$(\tau_k^0(x),\dots,\tau_k^{d+1}(x)) = (x_k^0,\dots,x_k^{d+1}).$$

The sequences $(\tau_k^0)_{k\in\mathbb{N}}$, $(\tau_k^1)_{k\in\mathbb{N}}$, $(\tau_k^i)_{k\in\mathbb{N}}$, $i = 2,\dots,d+2$, are independent and respectively Gaussian, exponential and uniform on $[-1,1]$. We let

$$E = \mathbb{R}\times]0,\infty[\times]-1,1[^{d-1},$$

$$\bar{E} = \mathbb{R}\times[0,\infty[\times[-1,1]^d,$$

and

$$B_+ = \{x \in B \ : \ x_k \in \bar{E},\ k \in \mathbb{N}\},$$

$$B_- = \{x \in B \ : \ x_k \in E,\ k \in \mathbb{N}\}.$$

The random configurations γ can be constructed as the sets of points

$$\gamma = \{T_k(x) \ : \ k \geq 1\} \subset \mathbb{R}_+ \times [-1,1]^d, \quad x \in B_+,$$

defined as

$$T_k(x) = \left(\sum_{i=0}^{i=k-1} \tau_i^1(x), \tau_k^2(x),\dots,\tau_k^{d+1}(x) \right), \quad x \in B, \quad k \geq 1.$$

On the other hand, it is well known that the classical Brownian motion on $[0,1]$ can be constructed as

$$W(t) = t\tau_0^0 + \sqrt{2}\sum_{n=1}^{\infty} \frac{\tau_n^0}{2n\pi}\sin(2n\pi t), \quad t \in [0,1],$$

i.e.,

$$\tau_n^0 = \sqrt{2}\int_0^1 \sin(2\pi nt)dW(t), \ n \geq 1, \quad \tau_0^0 = \int_0^1 dW(t) = W(1),$$

and if $(z(t))_{t\in\mathbb{R}_+}$ is an adapted process given as

$$z(t) = F(0,0) + \sqrt{2}\sum_{n=1}^{\infty} F(n,0)\cos(2n\pi t), \quad t \in [0,1],$$

then the stochastic integral of $(z(t))_{t\in\mathbb{R}_+}$ with respect to $(W(t))_{t\in\mathbb{R}_+}$ is written as

$$\int_0^1 z(t)dW(t) = \sum_{n=0}^{\infty} F(n,0)\tau_n^0,$$

and we have

$$\int_0^1 z^2(t)dt = \sum_{n=0}^{\infty}(F(n,0))^2.$$

Let also

$$E_i = \begin{cases} \mathbb{R}^{d+2}, & i = 0, \\ \{(y^0,\dots,y^{d+1}) \in \mathbb{R}^{d+2} \ : \ y^1 = 0\}, & i = 1, \\ \{(y^0,\dots,y^{d+1}) \in \mathbb{R}^{d+2} \ : \ y^i \in \{-1,1\}\}, & i = 2,\dots,d+1, \end{cases}$$

and

$$B_k^i = \{x \in B \ : \ x_k \in E_i\}, \quad k \in \mathbb{N}, \ i = 1,\dots,d+1.$$

We denote by $(e_k)_{k\geq 0}$ the canonical basis of $H = l^2(\mathbb{N},\mathbb{R}^{d+2}) = l^2(\mathbb{N})\otimes\mathbb{R}^{d+2}$, with

$$e_k = (e_k^0,\dots,e_k^{d+1}), \quad k \in \mathbb{N}.$$

In this framework, the shift of Brownian motion by a process $(\psi(s))_{s\in[0,1]}$ and the random diffeomorphism $\phi : \mathbb{R}_+ \times [-1,1]^d \longrightarrow \mathbb{R}_+ \times [-1,1]^d$ will be replaced by a random variable $F : B \longrightarrow H$ whose components are denoted by $(F(k,i))_{k\in\mathbb{N},i=0,\dots,d+1}$. The link between F and ψ, ϕ is the following:

$$F(k,0) = \begin{cases} \sqrt{2}\int_0^1 \sin(2\pi kt)\psi(t)dt, & k \geq 1, \\ \int_0^1 \psi(t)dt, & k = 0, \end{cases}$$

$$\tau_k^1 + F(k,1) = \phi^1(T_{k+1}) - \phi^1(T_k), \quad k \geq 0,$$

$$\tau_k^i + F(k,i) = \phi^i(T_k), \quad k \geq 0, \ i = 2,\dots,d+1.$$

4. Anticipating Girsanov theorem

In this section we will state the extension of the Girsanov theorem for Lévy processes to non-adapted shifts, and compare it to its classical adapted version (Th. 2.2). Before that we need to introduce the tools of gradient and divergence operator which will be used in the expressions of densities. Given X a real separable Hilbert space with orthonormal basis $(h_i)_{i\in\mathbb{N}}$, let $H\otimes X$ denote the completed Hilbert–Schmidt tensor product of H with X. Let $\mathcal{S}$ be the set of functionals on B of the form $f(\tau_{k_1},...,\tau_{k_n})$, where $n\in\mathbb{N}$, $k_1,...,k_n\in\mathbb{N}$, and f is a polynomial or $f\in\mathcal{C}_c^\infty(E^n)$. We define a set of smooth vector-valued functionals as

$$\mathcal{S}(X)=\left\{\sum_{i=0}^{i=n}F_ih_i\ :\ F_0,\dots,F_n\in\mathcal{S},\ h_0,\dots,h_n\in X,\ n\in\mathbb{N}\right\},$$

which is dense in $L^2(B,P;X)$.

Definition 4.1. *We define a gradient $D:\mathcal{S}(X)\to L^2(B,H\otimes X)$ by*

$$(DF(x),h)_{H\otimes X}=\lim_{\varepsilon\to 0}\frac{F(x+\varepsilon h)-F(x)}{\varepsilon},\qquad x\in B,\ h\in H.$$

The coordinates of $DF\in L^2(B;H\otimes X)$ are denoted by $(D_k^iF)_{(k,i)\in\mathbb{N}\times\{0,\dots,d+1\}}$. For $u\in\mathcal{S}(H\otimes X)$, we write

$$u=\sum_{k=0}^{\infty}\sum_{i=0}^{i=d-1}u_k^ie_k^i,\quad u_k\in\mathcal{S}(X),\quad k\in\mathbb{N}.$$

Let

$$\mathcal{U}(X)=\left\{u\in\mathcal{S}(H\otimes X)\ :\ u_k^i=0\text{ on }B_k^i,\ k\in\mathbb{N},\ i=0,\dots,d+1\right\},$$

which is dense in $L^2(B;H\otimes X)$.

Proposition 4.2. *The operator $D:L^2(B;X)\to L^2(B;H\otimes X)$ is closable and has an adjoint operator $\delta:\mathcal{U}(X)\to L^2(B;X)$, with*

$$E_P\left[(DF,u)_{H\otimes X}\right]=E\left[(\delta(u),F)_X\right],\quad u\in\mathcal{U}(X),F\in\mathcal{S}(X),\tag{4.1}$$

where δ is defined as

$$\delta(u)=\sum_{k\in\mathbb{N}}\tau_k^0u_k^0+u_k^1-\text{trace }D_ku_k,\quad u\in\mathcal{U}(X),$$

with

$$\text{trace }D_ku_k=D_k^0u_k^0+\dots+D_k^{d+1}u_k^{d+1},\quad u\in\mathcal{U}(X).$$

Proof. This result is proved by finite dimensional integration by parts with respect to λ_0, under the boundary conditions imposed on elements of $\mathcal{U}(X)$. □

Given a Hilbert–Schmidt operator $K : H \longrightarrow H$, the Carleman–Fredholm determinant of $I_H + K$ is defined as

$$\det_2(I_H + K) = \prod_{i=0}^{\infty}(1 + \alpha_i)\exp(-\alpha_i), \tag{4.2}$$

where $(\alpha_k)_{k\in\mathbb{N}}$ are the eigenvalues of K, counted with their multiplicities, cf. [6].

Theorem 4.3. *Let $F : B \to H$ be such that $h \mapsto F(x+h)$ is continuously differentiable in $H \otimes H$ on $\{h \in H \ : \ x + h \in B_+\}$, a.s. for $x \in B$. Assume that*

(i) $(I_B + F)(B_-) = B_-$,
(ii) $(I_B + F)(B_k^i) \subset B_k^i$, $k \in \mathbb{N}$, $i = 1, \dots, d+1$,
(iii) $I_B + F : B \longrightarrow B$ *is a.s. bijective,*
(iv) $I_H + DF : H \longrightarrow H$ *is a.s. invertible.*

Then

$$E_P[f] = E_P\left[|\det_2(I_H + DF)|\exp\left(-\delta(F) - \frac{1}{2}|\pi^0 F|_H^2\right) f \circ (I_B + F)\right]$$

for $f : B \longrightarrow \mathbb{R}$ measurable and bounded, where $\pi^0 : H \longrightarrow l^2(\mathbb{N})$ is the projection operator defined as $\pi^0(u) = (u_k^0)_{k\in\mathbb{N}}$.

This result is a particular case of Th. 4.5 stated below and proved in Sect. 6. The integrability condition (2.2) in Th. 2.1 is ensured by the hypothesis of Th. 4.3. We also make the following remarks:

- The boundary condition (*ii*) in Th. 4.3 is natural. For $i = 1$, it states that if two points in γ have the same jump times, then their images by $I_B + F$ also have the same jump times. For $i = 2, \dots, d+1$, it means that

$$\tau_k^i = \pm 1 \Rightarrow F(k, i) = 0,$$

  i.e., if a point lies at a boundary of $[-1, 1]^d$, then its image by $I_B + F$ lies at the same boundary.
- Let us check that in the adapted Poisson case ($F(n, 0) = 0$, $n \in \mathbb{N}$), the above result is in agreement with Th. 2.1 (ν is the flat intensity given by (3.1)). If $I_B + F : B \longrightarrow B$ satisfies the hypothesis of Th. 4.3 and corresponds to a smooth $\mathcal{F}_t$-predictable random diffeomorphism $\phi : \mathbb{R}_+ \times [-1, 1]^d \longrightarrow \mathbb{R}_+ \times [-1, 1]^d$ satisfying the hypothesis of Th. 2.1, then $D_k^i F(l, j) = 0$, $k > l$, hence $I_H + DF$ is a block diagonal matrix, each $d \times d$ diagonal block being equal to the Jacobian determinant $|\partial\phi(T_k)|$. We have $\tau_k^1 + F(k, 1) = \phi^1(T_{k+1}) - \phi^1(T_k)$, $k \geq 1$, $\phi^i(T_k) = \pm 1$ if $\tau_k^i = \pm 1$, $k \geq 1$, $i = 2, \dots d+1$, and

$$\begin{aligned}\sum_{k=0}^{\infty} F(k,1) &= \lim_{k\to\infty} (\phi^1(T_k) - T_k^1) \\ &= \lim_{k\to\infty} \nu(\phi([0,T_k]\times[-1,1]^d)) - \nu([0,T_k]\times[-1,1]^d) \\ &= \int_{\mathbb{R}_+\times[-1,1]^d} (|\partial\phi(s,x)| - 1)\nu(ds,dx), \quad a.s.\end{aligned}$$

Hence the formula of Th. 2.1:

$$|\det{}_2(I_H + DF)| \exp(-\delta(F)) = |\det(I_H + DF)| \exp\left(-\sum_{k=0}^{\infty} F(k,1)\right)$$

$$= \exp\left(-\sum_{k=0}^{\infty} F(k,1)\right) \prod_{k=1}^{\infty} |\partial\phi(T_k)|$$

$$= \exp\left(-\int_{\mathbb{R}_+\times[-1,1]^d} (|\partial\phi(s,x)| - 1)\nu(ds,dx)\right) \prod_{k=1}^{\infty} |\partial\phi(T_k)|.$$

- Still in the adapted Poisson case, the Carleman–Fredholm factorization $\det_2(I_H + DF)$ of the determinant has some similarity with the expression (2.5) of the density $L_\infty = dQ/dP$. The conditions $Z = (|\partial\phi| - 1) \in L^1(\mathbb{R}_+ \times [-1,1]^d, \tilde{\nu})$, resp. $Z(|\partial\phi| - 1) \in L^2(\mathbb{R}_+ \times [-1,1]^d, \tilde{\nu})$ are the respective analog of the trace class and Hilbert–Schmidt hypothesis on DF, a.s. More precisely we have

$$\det(I_H + DF) = \prod_{k=1}^{\infty} |\partial\phi(T_k)| \exp\left(-\sum_{i=1}^{d+1} D_k^i F(k,i)\right).$$

If (and only if) $d = 0$, then the Carleman–Fredholm determinant $\det_2(I_H + DF)$ coincides exactly with the factorization (2.6), i.e.,

$$\det{}_2(I_H + DF) = \prod_{k=1}^{\infty} |\phi'(T_k)| \exp(1 - \phi'(T_k)),$$

and $\delta(F) = \int_0^\infty \phi'(t)(\gamma(dt) - \nu(dt))$.
- In the general adapted case, i.e., if F also perturbs the Brownian component, then this type of result can not be checked via the above elementary computation. If $I_B + F$ satisfies all the above smoothness and adaptedness conditions, the comparison of Th. 4.3 to the classical Girsanov theorem Th. 2.1 yields an equality between

$$
\begin{aligned}
&|\det_2(I_H + DF)| \exp\left(-\delta(F) - \frac{1}{2}|\pi^0 F|_H^2\right) \\
&= |\det(I_H + DF)| \exp\left(-\sum_{k=0}^{\infty} \tau_k^0 F(k,0) - F(k,1) - \frac{1}{2}(F(k,0))^2\right) \\
&= \exp\left(-\int_0^1 z(s)dW(s) + \int_{\mathbb{R}_+\times[-1,1]^d} (|\partial\phi(s,x)| - 1)\nu(ds,dx) - \frac{1}{2}\int_0^1 z(s)^2 ds\right) \\
&\quad \times |\det(I_H + DF)|
\end{aligned}
$$

and

$$
\exp\left(-\int_0^1 z(s)dW(s) + \int_{\mathbb{R}_+\times[-1,1]^d} (|\partial\phi(s,x)| - 1)\nu(ds,dx) - \frac{1}{2}\int_0^1 z(s)^2 ds\right)
\times \prod_{k=1}^{\infty} |\partial\phi(T_k)|,
$$

hence the expression of the determinant in the case of an adapted transformation of a Brownian motion and a Poisson random measure satisfying the hypothesis of Th. 2.1 and Th. 4.3:

$$
\det(I_H + DF) = \prod_{k=1}^{\infty} |\partial\phi(T_k)|.
$$

We will prove a result which is more general than Th. 4.3 and does not require $I_B + F$ to be bijective. For this we need to consider the following class of transformations, cf. [10], [21].

Definition 4.4. *A random variable $F : B \to H$ is said to be $\mathcal{H}C^1_{loc}$ if there is a random variable $R : B \longrightarrow [0,\infty]$ with $R > 0$ a.s. such that $h \to F(x+h)$ is continuously differentiable in $H \otimes H$ on*

$$
\{h \in H \ : \ \|h\|_H < R(x) \text{ and } x + h \in B_+\},
$$

for any $x \in B_+$.

Our main result is the following, it is formulated for not necessarily invertible shifts, as in [21] on the Wiener space.

Theorem 4.5. *Let $F \in \mathcal{H}C^1_{loc}$ and $M = \{x \in B_+ \ : \det_2(I_H + DF) \neq 0\}$. Assume that*

(i) $(I_B + F)(B_-) \subset B_-$ *and*
(ii) $(I_B + F)(B^i_k) \subset B^i_k$, $k \in \mathbb{N}$, $i = 1, \ldots, d+1$.

Then

(i) $N(x; M) := card((I_B + F)^{-1}(x) \bigcap M)$, $x \in B$, *is at most countably infinite,*

(ii) *we have*

$$E_P[fN(\cdot;M)] = E_P\left[|\det{}_2(I_H+DF)|\exp\left(-\delta(F)-\frac{1}{2}|\pi^0F|_H^2\right) f\circ(I_B+F)\right]$$

for $f \in \mathcal{C}_b^+(B)$,

(iii) *the measure* $(I_B+F)_*(P_{|M})$ *is absolutely continuous with respect to* P, *and*

$$\frac{d(I_B+F)_*P_{|M}}{dP}(x) = \sum_{\theta\in(I_B+F)^{-1}(x)\cap M} \exp\left(\delta(F)(\theta)+\frac{1}{2}|\pi^0F(\theta)|_H^2\right)\frac{1}{|\det{}_2(I_H+DF(\theta))|}.$$

Proof. cf. Sect. 6. □

Th. 4.3 is a particular case of Th. 4.5 with $R=\infty$, P-a.s., $P(M)=1$ and I_B+F P-a.s. bijective.

5. Technical results

In this section, some further notation is introduced, and basic properties of D and δ are stated.

Definition 5.1. *For* $p \geq 1$, *we call*

- $\mathbb{D}_{p,1}(X)$ *the completion of* $\mathcal{S}(X)$ *with respect to the norm*

$$\|F\|_{\mathbb{D}_{p,1}(X)} = \|\|F\|_X\|_{L^p(B)} + \|\|DF\|_{H\otimes X}\|_{L^p(B)},$$

- $\mathbb{D}_{p,1}^{\mathcal{U}}(H)$ *the completion of* $\mathcal{U}(\mathbb{R})$ *with respect to the norm* $\|\cdot\|_{\mathbb{D}_{p,1}(H)}$,
- $\mathbb{D}_{\infty,1}(X)$, *resp.* $\mathbb{D}_{\infty,1}^{\mathcal{U}}(H)$ *the subset of* $\mathbb{D}_{2,1}(X)$, *resp.* $\mathbb{D}_{2,1}^{\mathcal{U}}(H)$ *made of the random variables* F *for which* $\|F\|_{\mathbb{D}_{\infty,1}(X)}$, *resp.* $\|F\|_{\mathbb{D}_{\infty,1}(H)}$ *is bounded.*

For $p \in [1,\infty]$, *we call*

- $\mathbb{D}_{p,1}^{loc}(X)$, *resp.* $\mathbb{D}_{p,1}^{\mathcal{U},loc}(H)$, *the sets of functionals* F *such that there is a measurable almost sure partition* $(A_n)_{n\in\mathbb{N}}$ *of* B *and* $F_n \in \mathbb{D}_{p,1}(X)$, *resp.* $F_n \in \mathbb{D}_{p,1}^{\mathcal{U}}(X)$, *with* $F_n = F$ *a.s. on* A_n, $n \in \mathbb{N}$.

For $p = 2$,

- *let* $\mathrm{Dom}(\delta;X)$ *denote the domain of the closed extension of* δ.

The interest in the space $\mathbb{D}_{2,1}^{\mathcal{U}}(H)$ is that it is a Hilbert space contained in $\mathrm{Dom}(\delta;X)$, as shown in the following proposition.

Proposition 5.2. (i) *The operator* δ *is continuous from* $\mathbb{D}_{2,1}^{\mathcal{U}}(H)$ *into* $L^2(B)$ *with*

$$\|\delta(F)\|_{L^2(B)}^2 \leq (d+2)\|F\|_{\mathbb{D}_{2,1}^{\mathcal{U}}(H)}^2, \quad F \in \mathbb{D}_{2,1}^{\mathcal{U}}(H). \tag{5.1}$$

(ii) *The operators D and δ are local, cf. [2], [12], [14], i.e., for $F \in \mathbb{D}_{2,1}(X)$, resp. $F \in \mathrm{Dom}(\delta; X)$, we have $DF = 0$ a.s. on $\{F = 0\}$, resp. $\delta(F) = 0$ a.s. on $\{F = 0\}$.*

Proof. (i) Let $F \in \mathcal{U}(\mathbb{R})$. We have

$$\delta(F) = \sum_{k=0}^{\infty} \left(\tau_k^0 F(k,0) + F(k,1) - \sum_{i=0}^{d+1} D_k^i F(k,i) \right),$$

and

$$(\delta(F))^2 \leq (d+2) \left(\sum_{k=0}^{\infty} \tau_k^0 F(k,0) - D_k^0 F(k,0) \right)^2$$
$$+(d+2) \left(\sum_{k=0}^{\infty} F(k,1) - D_k^1 F(k,1) \right)^2 + (d+2) \sum_{i=2}^{d+1} \left(\sum_{k=0}^{\infty} D_k^i F(k,i) \right)^2,$$

hence from the Gaussian, exponential and uniform cases, cf. [17], [14], [15], we have

$$\|\delta(F)\|^2_{L^2(B)} \leq (d+2) E_P \left[\sum_{k=0}^{\infty} (F(k,0))^2 \right]$$
$$+(d+2) E_P \left[\sum_{k,l=0}^{\infty} (D_k^0 F(l,0))^2 + (D_k^1 F(l,1))^2 + \sum_{i=2}^{d+1} (D_k^i F(l,i))^2 \right]$$
$$\leq (d+2) \|\pi^0 F\|^2_{\mathbb{D}^{\mathcal{U}}_{2,1}(H)}.$$

(See Th.4.3 for the definition of π^0.) The proof of (ii) relies only on the duality relation between D and δ and on the density of $\mathcal{U}(X)$ in $L^2(B; H \otimes X)$. □

The proof of the following result is directly adapted from [4], [11], [13], [14]; it remains valid by replacing λ_0 with any probability measure on $\mathbb{R}^{d+2}$. Let $\mathcal{F}_n$ denote the σ-algebra generated by $\tau_0, \ldots, \tau_n$.

Lemma 5.3. *Let $F \in L^2(B; X)$ and $F_n = E[F \mid \mathcal{F}_n] \in \mathbb{D}_{2,1}(X)$, $n \in \mathbb{N}$.*

- *$F \in \mathbb{D}_{2,1}(X)$ if and only if $F_n \in \mathbb{D}_{2,1}(X)$ $\forall n \in \mathbb{N}$ and $(DF_n)_{n \in \mathbb{N}}$ converges in $L^2(B; H \otimes X)$. In this case,*

$$\|DF_n\|_{H \otimes X} \leq \|DF\|_{H \otimes X}, \quad a.s., \ n \in \mathbb{N}.$$

- *F_n belongs to $\mathbb{D}_{2,1}(\mathbb{R})$ if and only if there exists*

$$f \in W^{2,1}(E^{n+1}, \lambda_0^{\otimes n+1})$$

such that $F_n = f(\tau_0, \ldots, \tau_n)$. In this case, $DF_n = (\partial_k f(\tau_0, \ldots, \tau_n))_{k \in \mathbb{N}}$.

- *Assume that for some $c > 0$,*

$$\|F(x+h) - F(x)\|_X \leq c \|h\|_H, \quad h \in H, \ x \in B_+, \ x+h \in B_+.$$

Then $F \in \mathbb{D}_{2,1}(X)$ and $\|DF\|_{H \otimes X} \leq c$, a.s.

We denote by π_n the application $\pi_n : B \longrightarrow H$ defined by $\pi_n(x) = \left(x_k 1_{\{k\leq n\}}\right)_{k\in\mathbb{N}}$.

Lemma 5.4. *Let $F : B \to H$ be measurable and bounded such that*

(i) $(I_B + F)(B_k^i) \subset B_k^i$, $k \in \mathbb{N}$, $i = 1, \ldots, d+1$,

(ii) *F is Lipschitz on B_+ with Lipschitz constant $c > 0$:*

$$\|F(x+h) - F(x)\|_H < c\|h\|_H, \quad h \in H, \ x \in B_+, \ x + h \in B_+.$$

Then $F \in \mathbb{D}^{\mathcal{U}}_{\infty,1}(H)$, and there is a sequence $(F_n)_{n\in\mathbb{N}} \subset \mathcal{U}(\mathbb{R})$ that converges in $\mathbb{D}^{\mathcal{U}}_{2,1}(H)$ to F with

(i) $\| \|F_n\|_H \|_\infty \leq \| \|F\|_H \|_\infty$,

(ii) $\| \|DF_n\|_{H\otimes H} \|_\infty \leq c$, $n \in \mathbb{N}$.

Proof. Let

$$\tilde{E}_i = \begin{cases} E, & i = 0, \\ \mathbb{R} \times \mathbb{R} \times [-1,1]^d, & i = 1, \\ \mathbb{R} \times \mathbb{R}_+ \times [-1,1]^{i-2} \times \mathbb{R} \times [-1,1]^{d-i+1}, & i = 2, \ldots, d+1, \end{cases}$$

and

$$\tilde{B}_k^i = \{x \in B_+ \ : \ x_k \in \tilde{E}_i\}, \quad i = 0, \ldots, d+1, \ k \in \mathbb{N}.$$

First we note that after putting $F = 0$ on B_+^c, the Lipschitz condition on F extends to $\tilde{B}_k^i$ as

$$|F(k,i)(x+h) - F(k,i)(x)| < c\|h\|_H, \quad h \in H, \ x \in \tilde{B}_k^i, \ x + h \in \tilde{B}_k^i.$$

Let $F_n = \pi_n E[F \mid \mathcal{F}_n]$, $n \in \mathbb{N}$. The sequence $(F_n)_{n\in\mathbb{N}}$ converges to F in $\mathbb{D}_{2,1}(H)$ and satisfies (i) and (ii). There exists a function $f_k^i \in W^{2,1}(\mathbb{R}^{(d+2)(n+1)}, dx)$ which has a Lipschitz version on $\mathbb{R}^{(d+2)(n+1)}$ and support in $\bar{E}^{n+1}$, such that $F_n(k,i) = f_k^i(\tau_0, \ldots, \tau_n)$ P-a.e., $k = 0, \ldots, n$, $i = 0, \ldots, d+1$. Let $\Psi \in \mathcal{C}_c^\infty(\mathbb{R}^{(d+2)(n+1)})$ with support in $[-2,0]^{k(d+2)+i} \times [0,2] \times [-2,0]^{(n-k)(d+2)+(d+2-i)}$, $0 \leq \Psi \leq 1$ and $\int_{\mathbb{R}^{(d+2)(n+1)}} \Psi(x)dx = 1$. Let for $N \geq 2$ and $y \in \bar{E}^{n+1}$:

$$\phi_{k,i,N}(y) = \begin{cases} \left(\frac{1}{N}\right)^{(d+2)(n+1)} \int_{\tilde{E}_i} \Psi(N(y-x)) f_k^i(x)dx, \ i = 0, 1, \\ \left(\frac{1}{N}\right)^{(d+2)(n+1)} \int_{\tilde{E}_i} \Psi(N(y-x)) f_k^i(x)dx, \ y_k^i < 0, \ i = 2, \ldots, d+1, \\ \left(\frac{1}{N}\right)^{(d+2)(n+1)} \int_{\tilde{E}_i} \Psi(N(y+x)) f_k^i(x)dx, \ y_k^i \geq 0, \ i = 2, \ldots, d+1. \end{cases}$$

For $k \in \mathbb{N}$ and $i = 0, \ldots, d+1$, let $G_N(k,i) = \phi_{k,i,N}(\tau_0, \ldots, \tau_n)$, $k = 0, \ldots, n$, and $G_N(k,i) = 0$, $k > n$. Then $G_N \in \mathcal{U}(\mathbb{R})$, $N \geq 2$, and $(G_N)_{N\geq 2}$ converges to F_n in $\mathbb{D}_{2,1}(H)$ and satisfies (i) and (ii). □

We refer to [10] for the following definition on the Wiener space.

Definition 5.5. *If $A \subset B$ is measurable, let*

$$\rho_A(x) = \begin{cases} \inf_{h \in H} \{ \|h\|_H \ : \ x + h \in A \}, & x \in B, \\ \infty, & x \notin A + H, \ x \in B. \end{cases}$$

Let $\phi \in \mathcal{C}_c^\infty(\mathbb{R})$ with $\|\phi\|_\infty \leq 1$, such that $\phi = 0$ on $[2/3, \infty[$, $\phi = 1$ on $[0, 1/3]$ and $\|\phi'\|_\infty < 4$. If A is σ-compact, then

$$\|\phi(\rho_A(x+h)) - \phi(\rho_A(x))\|_H \leq \|\phi'\|_\infty \|h\|_H, \quad x \in B, \ h \in H,$$

hence $\phi(\rho_A) \in \mathbb{D}_{\infty,1}(\mathbb{R})$ with $\|D\phi(\rho_A)\|_H \leq \|\phi'\|_\infty$.

Lemma 5.6. *Let $n_0 \in \mathbb{N}$ and $F \in \mathcal{H}C^1_{loc}$, such that*

$$(I_B + F)(B^i_k) \subset B^i_k, \quad i = 1, \ldots, d+1, \quad k > n_0.$$

Let $\alpha, \beta > 0$ and

$$A = \Big\{ x \in B_- \ : \quad \begin{aligned} & \rho_{B^i_k}(\pi_{n_0} x) > 4/\alpha, \ k \in \mathbb{N}, \ i = 1, \ldots, d+1, \\ & R(x) \geq 4/\alpha, \\ & \sup_{\|h\|_H \leq 2/\alpha} \|F(x+h)\|_H \leq \beta/(6\alpha), \\ & \sup_{\|h\|_H \leq 2/\alpha} \|DF(x+h)\|_{H \otimes H} \leq \beta/6 \Big\} \end{aligned}$$

and $\tilde{F} = \phi(\alpha \rho_G) F$, where G is a σ-compact set contained in A. Then

$$\|\tilde{F}(x+h) - \tilde{F}(x)\|_H \leq (5\beta/6)\|h\|_H, \quad h \in H, \ x, x+h \in B_+,$$

and $\| \|\tilde{F}\|_H \|_\infty \leq \beta/(6a)$. Consequently $\tilde{F} \in \mathbb{D}^{\mathcal{U}}_{\infty,1}(H)$, and $F \in \mathbb{D}^{\mathcal{U},loc}_{\infty,1}(H)$.

Proof. Any $x \in A$ satisfies $x^1_k > 4/\alpha$, $1 - x^i_k > 4/\alpha$, and $x^i_k + 1 > 4/\alpha$, $i = 2, \ldots, d+1$, $k \leq n_0$. For x in B^i_k we have $\rho_A(x) \geq 4/\alpha$, hence $\phi(\alpha \rho_A(x)) = 0$, and $(I_B + \tilde{F}(k,i))(B^i_k) \subset B^i_k$, $0 \leq k \leq n_0$, $i = 1, \ldots, d+1$. From Lemma 5.4 it follows that $\tilde{F} \in \mathbb{D}^{\mathcal{U}}_{\infty,1}(H)$. The fact that $F \in \mathbb{D}^{\mathcal{U},loc}_{\infty,1}(H)$ is proved by covering B with a countable collection of sets such as A, with $\alpha, \beta \in \mathbb{Q} \cap]0, \infty[$. □

Lemma 5.7. *Let $F, G \in \mathcal{S}(H)$ and $T = I_B + F$. We have $G \circ T \in \mathrm{Dom}(\delta)$ and*

$$\delta(G) \circ T = \delta(G \circ T) + \mathrm{trace}\,(DF^*(DG) \circ T) + (\pi^0 F, G \circ T)_H.$$

Proof. We have $\delta(G \circ T) \in \mathcal{S}$ and

$$\begin{aligned}
\delta(G \circ T) &= \sum_{k=0}^{\infty} \tau_k^0 G(k,0) \circ T + G(k,1) \circ T - \sum_{k=0}^{\infty} \sum_{i=0}^{d+1} D_k^i \left(G(k,i) \circ T\right) \\
&= \sum_{k=0}^{\infty} \tau_k^0 G(k,0) \circ T + G(k,1) \circ T - \sum_{k=0}^{\infty} \sum_{i=0}^{i=d+1} D_k^i (I_B + F)^* (DG(k,i)) \circ T \\
&= -(\pi^0 F, G \circ T)_H + \delta(G) \circ T - \sum_{i,j=0}^{i=d+1} \sum_{k,l=0}^{\infty} D_k^i F(l,j) (D_l^j G(k,i)) \circ T \\
&= \delta(G) \circ T - \mathrm{trace}\, (DF^*(DG) \circ T) - (\pi^0 F, G \circ T)_H.
\end{aligned}$$

□

6. Proof of Theorem 4.5

Since from Prop. 5.2, $\mathbb{D}_{2,1}^{\mathcal{U}}(H) \subset \mathrm{Dom}(\delta)$, we define

$$\Lambda_F = \det{}_2(I_H + DF) \exp\left(-\delta(F) - \frac{1}{2}|\pi^0 F|_H^2\right), \quad F \in \mathbb{D}_{2,1}^{\mathcal{U},loc}(H). \tag{6.1}$$

The proof of Th. 4.5 is done in two main steps: first we treat the case of Lipschitz transformations in the following lemma. Then we use the fact that $F \in \mathcal{H}C_{loc}^1$ can be locally written as a Lipschitz transformation, as in [10]. Let $P_n^\perp$ denote the image measure of P by $\pi_n^\perp := I_B - \pi_n$.

Proposition 6.1. *Let $n_0 \in \mathbb{N}$, let $K : H \to \pi_{n_0} H$ be a linear operator such that $I_H + K$ is invertible, and let $v \in \pi_{n_0} H$. Let $F : B \to H$ be measurable, bounded with bounded support, such that*

(i) $(I_B + F)(B_k^i) \subset B_k^i$, $k \in \mathbb{N}$, $i = 1, \ldots, d+1$,
(ii) $\pi_{n_0}^\perp \circ (I_B + F)(B) \subset B_+$, *a.s.*,
(iii) *F is Lipschitz on B_- with Lipschitz constant $c < \left(\|(I_H + K)^{-1}\|_\infty\right)^{-1}$:*

$$\|F(x+h) - F(x)\|_H \le c\|h\|_H, \quad h \in H, \ x \in B_-, \ x + h \in B_-. \tag{6.2}$$

Then $I_B + F + K + v$ is injective and there is $r > 0$ such that

$$E_P[f] = \int_B |\Lambda_{F+K+v}| \, f \circ (I_B + F + K + v) \, d\lambda_r^{\otimes(n_0+1)} \otimes P_{n_0}^\perp,$$

for $f : B \longrightarrow \mathbb{R}$ bounded measurable with support in B_-.

Proof. Step 1: finite dimensional case. The injectivity of $I_B + F + K + v$ follows from (6.2) as in [10], [14]. Let $F = 0$ on B_+^c. Let $(F_n)_{n \ge n_0} \subset \mathcal{U}(\mathbb{R})$ be the sequence given by Lemma 5.4, converging to F in $\mathbb{D}_{2,1}(H)$, and let $T_n = I_B + F_n + K + v$. Since $\pi_{n_0}^\perp (I_B + F)(B) \subset B_+$, by construction the sequence $(F_n)_{n \in \mathbb{N}}$ also satisfies $\pi_{n_0}^\perp (I_B + F_n)(B) \subset B_+$, $n \in \mathbb{N}$. Replicating the argument of [10], [14], [21], we show that $I_B + F_n \circ (I_B + K)^{-1} + v$ is contractive from (6.2), hence bijective on B with inverse $I_B + G_n = (I_B + K) \circ T_n^{-1}$, where G_n satisfies

$$G_n = -F_n \circ (I_B + K)^{-1} \circ (I_B + G_n) - v, \tag{6.3}$$

and

$$\|DG_n\|_{H \otimes H} \leq c\|(I_H + K)^{-1}\|_\infty / (1 - c\|(I_H + K)^{-1}\|_\infty). \tag{6.4}$$

From (6.3) and the uniform boundedness in n and x of $(F_n)_{n \geq n_0}$ and $(G_n)_{n \geq n_0}$, there exists $r > 0$ such that $|T_n^{-1}(k,i)| < 1 + r$, $i = 2, \ldots, d+1$, and $T_n^{-1}(k,1) > -r$, $n \in \mathbb{N}$, $k = 0, \ldots, n_0$, on B_+. Let $g \in \mathcal{C}^\infty(\mathbb{R}^{(d+2)(n+1)}, \pi_{n+1}H)$ such that $F_n + K + v = g(\tau_0, \ldots, \tau_n)$, $n \geq n_0$. Since $\pi_{n_0}^\perp(I_B + F_n)(B) \subset B_+$ and $F_n = 0$ on B_-, $n \in \mathbb{N}$, from (6.3) we have

$$T_n(\{x \in B \ : \ x_k \in \bar{E}, \quad k > n_0\}) = \{x \in B \ : \ x_k \in \bar{E}, \quad k > n_0\}, \tag{6.5}$$

hence $x \mapsto x + g(x)$ is a diffeomorphism of $(\mathbb{R}^{d+2})^{n_0} \times \bar{E}^{n-n_0}$. The Jacobi theorem in dimension $(d+2)(n+1)$ gives:

$$\begin{aligned}
&\int_B |\Lambda_{F_n+K+v}| \, f(\pi_n \circ T_n) \, d\lambda_r^{\otimes(n_0+1)} \otimes P_{n_0}^\perp \\
&= \frac{1}{2^{(d-1)(n+1)}} \int_{\mathbb{R}^{(d+2)(n+1)}} 1_{E^{n+1}}(x + g(x)) f(x + g(x)) \\
&\qquad |\det(I_{\mathbb{R}^{n+1}} + \partial g)| \exp\left(- \sum_{k=1}^{k=n} g_k^i + x_k^i + x_k^i g_k^0 + \frac{1}{2}(x_k^0)^2 + \frac{1}{2}(g_k^0)^2 \right) m(dx) \\
&= \frac{1}{2^{(d-1)(n+1)}} \int_{\mathbb{R}^{(d+2)(n+1)}} 1_{E^{n+1}}(y) f(y) \exp\left(- \sum_{k=1}^{k=n} y_k^1 + \frac{1}{2}(y_k^0)^2 \right) m(dy) \\
&= E_P\left[f \circ \pi_n\right],
\end{aligned} \tag{6.6}$$

where $m(dx)$ denotes the Lebesgue measure on $\mathbb{R}^{(d+2)(n+1)}$, for $f \in \mathcal{C}_b^+(\mathbb{R}^{(d+2)(n+1)})$ with support in E^{n+1}. This relation extends to $f \in \mathcal{C}_b^+(B)$ with support in B_-:

$$\int_B |\Lambda_{F_n+K+v}| \, f \circ T_n \, d\lambda_r^{\otimes(n_0+1)} \otimes P_{n_0}^\perp = E_P\left[f\right].$$

Step 2: uniform integrability argument. From the de la Vallée–Poussin Lemma we need to find a bound on

$$\int_B 1_{B_-} \circ T_n |\Lambda_{F_n+K+v} \log |\Lambda_{F_n+K+v}|| \, d\lambda_r^{\otimes(n_0+1)} \otimes P_{n_0}^\perp,$$

uniformly on $n > n_0$. Since $(\|DF_n(x)\|_{H \otimes H})_{n \in \mathbb{N}}$ is bounded uniformly in $n \in \mathbb{N}$ and $x \in B$, $(|\det_2 DT_n(x)|)_{n \in \mathbb{N}}$ is uniformly lower and upper bounded from

$$|\det_2(DT_n(x))| \leq (1 + \|DT_n(x) - I_H\|_{H \otimes H}) \exp(1 + \|DT_n(x) - I_H\|_{H \otimes H}^2),$$

we only need to estimate

$$\begin{aligned}
&\int_B 1_{B_-} \circ T_n |\delta(F_n + K + v)\Lambda_{F_n+K+v}| \, d\lambda_r^{\otimes(n_0+1)} \otimes P_{n_0}^\perp \\
&= E_P\left[|\delta(F_n + K + v) \circ T_n^{-1}|\right] \\
&\leq E_P\left[|\delta(\pi_{n_0} F_n + K + v) \circ T_n^{-1}|\right] \qquad (6.7)\\
&\quad + E_P\Big[\Big|\text{trace}\Big[\big(D\,(-K \circ (I_B + K)^{-1} \\
&\qquad\qquad + (I_B + K)^{-1} \circ G_n)\big)^* \cdot \big(D\pi_{n_0}^\perp F_n\big) \circ T_n^{-1}\Big]\Big|\Big] \\
&\quad + E_P\left[|(\pi^0(G + G \circ (I_B + K)), \pi_{n_0}^\perp F_n \circ T_n^{-1})_H|\right] \\
&\quad + E_P\left[|\delta(\pi_{n_0}^\perp F_n \circ T_n^{-1})|\right], \qquad (6.8)
\end{aligned}$$

from (6.6) and Lemma 5.7, since

$$\begin{aligned}
T_n^{-1} &= (I_B + K)^{-1}(I_B + G_n) = (I_B + K)^{-1} + (I_B + K)^{-1}(I_B + G_n) \\
&= (I_B + K)(I_B + K)^{-1} - K(I_B + K)^{-1} + (I_B + K)^{-1}(I_B + G_n) \\
&= I_B - K(I_B + K)^{-1} + (I_B + K)^{-1}(I_B + G_n), \quad n \in \mathbb{N}.
\end{aligned}$$

The first three terms in (6.8) are uniformly bounded in n from (6.4). From (6.3), we have $\pi_{n_0}^\perp G_n = -\pi_{n_0}^\perp F_n \circ T_n^{-1}$, and $\pi_{n_0}^\perp G_n \in \mathcal{U}(\mathbb{R})$, hence from (5.1),

$$\begin{aligned}
E_P\left[|\delta(\pi_{n_0}^\perp F_n \circ T_n^{-1})|\right] &= E_P\left[|\delta(\pi_{n_0}^\perp G_n)|\right] \leq E_P\left[|\delta(\pi_{n_0}^\perp G_n)|^2\right] \\
&\leq (d+2) E_P\left[\|D\pi_{n_0}^\perp G_n\|_{H\otimes H}^2\right] + (d+2)\|F\|_\infty^2 \\
&\leq (d+2)(c\|(I_H + K)^{-1}\|_\infty / (1 - c\|(I_H + K)^{-1}\|_\infty))^2 \\
&\quad + (d+2)\|F\|_\infty^2, \quad n \geq n_0,
\end{aligned}$$

from (5.1). Choosing a subsequence we have the $\lambda_r^{\otimes(n_0+1)} \otimes P_{n_0}^\perp$-a.e. convergence of $(|\Lambda_{F_n+K+v}| f \circ T_n)_{n\geq n_0}$ to $|\Lambda_{F+K+v}| \; f \circ T$, and by uniform integrability

$$\int_B |\Lambda_{F+K+v}| \; f \circ T \; d\lambda_r^{\otimes(n_0+1)} \otimes P_{n_0}^\perp = E_P[f]. \qquad \square \qquad (6.9)$$

Proof. of Th. 4.5. We construct a family of sets that form a partition of M, such that F is Lipschitz and satisfies the hypothesis of Prop. 6.1 on each of those sets. Given $K : H \to \pi_{n_0} H$ a linear operator, $v \in \pi_{n_0} H$, $n > 8$, let

$$\begin{aligned}
A(n_0, n, K, v) = \Big\{ x \in B_- : \; & \rho_{B_k^i}(\pi_{n_0} x) > \tfrac{8}{n}, \;\; k \in \mathbb{N}, \; i = 1, \ldots, d+1, R(x) > \tfrac{4}{n}, \\
& \sup_{\|h\|_H \leq 1/n} \|F(x+h) - K(x+h) - v\|_H < \tfrac{1}{6n}\left(\|(I_H + K)^{-1}\|_\infty\right)^{-1}, \\
& \sup_{\|h\|_H \leq 1/n} \|DF(x+h) - K\|_{H\otimes H} < \tfrac{1}{n}\left(\|(I_H + K)^{-1}\|_\infty\right)^{-1} \Big\}.
\end{aligned}$$

Let $F_{K,v} = \phi(n\rho_{\tilde{A}(n_0,n,K,v)})(F - K - v)$, where $\tilde{A}(n_0, n, K, v)$ is a σ-compact modification of $A(n_0, n, K, v) \bigcap M$. Then from Lemma 5.6, $F_{K,v}$ and $\tilde{A}(n_0, n, K, v)$ satisfy the hypothesis of Prop. 6.1, and since $T = I_B + F_{K,v} + K + v$ on $\tilde{A}(n_0, n, K, v) \subset B_-$, we have from Prop. 6.1:

$$E_P\left[1_{\tilde{A}(n_0,n,K,v)} \; |\Lambda_F| \; f \circ T\right] = E_P\left[1_{T(\tilde{A}(n_0,n,K,v))} f\right].$$

Finally we deal with the non-invertibility of $T = I_B + F$ as in [21]. Denote by $(\tilde{A}_k)_{k\in\mathbb{N}}$ the countable family $(\tilde{A}(n_0, n, K, v))_{n_0,n,K,v}$ obtained by letting K, resp. v, run in the finite rank linear operators and vectors with rational coefficients. Let $M_n = \tilde{A}_n \bigcap \left(\bigcup_{i=0}^{i=n-1} \tilde{A}_i\right)^c$, $n \in \mathbb{N}^*$. We have the partition $\bigcup_{n\in\mathbb{N}^*} M_n = M$, and

$$\begin{aligned} E_P\left[|\Lambda_F|\, f\circ T\right] &= \sum_{n=0}^{\infty} E_P\left[1_{M_n}|\Lambda_F|\, f\circ T\right] \\ &= \sum_{n=0}^{\infty} E_P\left[1_{T(M_n)} f\right] = E_P\left[f N(x;M)\right]. \end{aligned}$$

The computation of $d(I_B+F)_*P_{|M}/dP$ follows from

$$E_P\left[1_M f\circ T\right] = \sum_{n=0}^{\infty} E_P\left[1_{T(M_n)}\frac{f}{\Lambda_F\circ T}\right] = E_P\left[f \sum_{\theta\in T^{-1}(x)\bigcap M} \frac{1}{\Lambda_F(\theta)}\right].$$

□

References

[1] S. Albeverio, Yu. G. Kondratiev, and M. Röckner. Analysis and geometry on configuration spaces. *J. Funct. Anal.*, 154(2):444–500, 1998.

[2] N. Bouleau and F. Hirsch. Formes de Dirichlet générales et densité des variables aléatoires réelles sur l'espace de Wiener. *J. Funct. Anal.*, 69:229–259, 1986.

[3] M. Brown. Discrimination of Poisson processes. *Ann. Math. Statist.*, 42:773–776, 1971.

[4] R. Buckdahn. Anticipative Girsanov transformations. *Probab. Theory Related Fields*, 91:211–238, 1992.

[5] R.H. Cameron and W.T. Martin. Transformations of Wiener integrals under translations. *Ann. of Math. (2)*, 45:386–396, 1944.

[6] N. Dunford and J. T. Schwartz. *Linear Operators*, volume II. Interscience publishers, New York, 1957.

[7] I.V. Girsanov. On transformating a certain class of stochastic processes by absolutely continuous substitution of measures. *Theory of Probability and Applications*, 5:285–301, 1960.

[8] J. Jacod. *Calcul stochastique et problèmes de martingales*, volume 714 of *Lecture Notes in Mathematics*. Springer Verlag, 1979.

[9] J. Jacod and J. Mémin. Caractéristiques locales et conditions de continuité absolue pour les semi-martingales. *Z. Wahrscheinlichkeitstheorie und Verw. Gebiete*, 35(1):1–37, 1976.

[10] S. Kusuoka. The nonlinear transformation of Wiener measure on Banach space and its absolute continuity. *Journal of the Faculty of Science of Tokyo University, Section IA, Mathematics*, 29:567–598, 1982.

[11] D. Nualart. Markov fields and transformations of the Wiener measure. In Tom Lindstrøm, Bernt Øksendal, and A. Süleyman Üstünel, editors, *The Proceedings of the Fourth Oslo-Silivri Workshop on Stochastic Analysis*, volume 8 of *Stochastics Monographs*, Oslo, 1993. Gordon and Breach.

[12] D. Nualart and E. Pardoux. Stochastic calculus with anticipative integrands. *Probab. Theory Related Fields*, 78:535–582, 1988.

[13] D. Ocone. A guide to the stochastic calculus of variations. In H. Körezlioğlu and A.S. Üstünel, editors, *Stochastic Analysis and Related Topics*, volume 1316 of *Lecture Notes in Mathematics*, Silivri, 1988. Springer-Verlag.

[14] N. Privault. Girsanov theorem for anticipative shifts on Poisson space. *Probab. Theory Related Fields*, 104:61–76, 1996.

[15] N. Privault. Absolute continuity in infinite dimensions and anticipating stochastic calculus. *Potential Analysis*, 8(4):325–343, 1998.

[16] N. Privault. Hypothesis testing and Skorokhod stochastic integration. *J. Appl. Probab.*, 37(2):560–574, 2000.

[17] R. Ramer. On nonlinear transformations of Gaussian measures. *J. Funct. Anal.*, 15:166–187, 1974.

[18] A.V. Skorokhod. On the differentiability of measures which correspond to stochastic processes. *Teor. Verojatnost. i. Primenen.*, 2:418–444, 1957.

[19] A.S. Üstünel and M. Zakai. Transformation of Wiener measure under anticipative flows. *Probab. Theory Related Fields*, 93:91–136, 1992.

[20] A.S. Üstünel and M. Zakai. Transformation of the Wiener measure under non-invertible shifts. *Probab. Theory Related Fields*, 99:485–500, 1994.

[21] A.S. Üstünel and M. Zakai. *Transformation of measure on Wiener space.* Springer Monographs in Mathematics. Springer-Verlag, Berlin, 2000.

[22] Y. Takahashi. Absolute continuity of Poisson random fields. *Publ. RIMS Kyoto University*, 26:629–647, 1990.

[23] M. Zakai and O. Zeitouni. When does the Ramer formula look like the Girsanov formula? *Ann. Probab.*, 20(3):1436–1440, 1992.

Départment de Mathématiques
Université deLa Rochelle
17072 La Rochelle Codex 1
France

Progress in Probability, Vol. 53, 203–205

Gaussian Measure of the Intersection of Two Absolutely Convex Sets

Ali Süleyman Üstünel

Abstract. Let (W, H, μ) be a finite dimensional Wiener space and let A, B be two measurable, convex, symmetric subsets of W such that $A \subset T^{-1}(A)$ for any contraction T. Then one has $\mu(A \cap B) \geq \mu(A)\,\mu(B)$.

Let (W, H, μ) be an abstract Wiener space. This means that W is a separable Fréchet space, μ is a Gaussian measure on W, with zero mean and unit covariance. H denotes the reproducing kernel Hilbert space associated to μ and is called the Cameron–Martin space. Due to the quasi-invariance of μ, the Gateaux derivative in the direction of H is well defined and closable in $L^p(\mu)$, for any $p \geq 1$; we shall denote its closure by ∇. The corresponding scale of Sobolev spaces is denoted by $\mathbb{D}_{p,k}$, where k is the differentiability index (cf. [4, 5]). The Sobolev space for vector valued random variables will be denoted by $\mathbb{D}_{p,k}(M)$, for those which take their values in the space M. We shall denote by $(P_t, t \geq 0)$ the Ornstein–Uhlenbeck semigroup on W, which is defined by

$$P_t f(x) = \int_W f(e^{-t}x + \sqrt{1 - e^{-2t}}y)\mu(dy)\,.$$

This semigroup possesses all the nice properties of the finite dimensional mollifiers, and is frequently used in the infinite dimensional problems. The adjoint of the operator ∇ with respect to μ will be denoted by δ, it is called the divergence operator and it coincides with the Ito integral on the adapted vector functions.

In this note we shall prove the following claim using an idea taken from [1]:

Theorem 1. *Let A and B be two convex, symmetric Borel subsets of W, where W is supposed to be finite dimensional. Assume that one of them, say A, has the property that $1_A \circ T \leq 1_A$ almost surely, for any contraction, then*

$$\mu(A \cap B) \geq \mu(A)\,\mu(B)\,. \tag{1}$$

Remark: This claim, without any restriction on the absolutely convex set A, was conjectured in the 1960s (cf. [3] and the references there). It is trivial for $n = 1$, it has been proven in [3] for $n = 2$ and it is unknown for $n \geq 3$.

Proof: First we prove the theorem in the finite dimensional case: assume first that $W = \mathbb{R}^n$. Let $g = P_\varepsilon 1_B$, where P_ε is the Ornstein–Uhlenbeck semigroup:

$$P_\varepsilon 1_B(x) = \int_W 1_B(e^{-\varepsilon}x + \sqrt{1 - e^{-2\varepsilon}}y)\mu(dy)\,.$$

Then g is a bounded log-concave and symmetric function from a theorem of Prékopa (cf. [2]), moreover its derivative ∇g is essentially bounded (the bounds explode as $\varepsilon \to 0$). Besides, it is strictly positive due to the positivity improving property of the Ornstein–Uhlenbeck semi-group. We can and we shall assume that $E[g] = 1$ by dividing g with $\mu(B)$. Let $\sigma(x)$ be the vector field defined by

$$\sigma(x) = \int_0^\infty e^{-t} P_t \nabla g(x) dt\,. \tag{2}$$

Note that $\delta\sigma(x) = g(x) - 1$ (cf. [4]), where δ denotes the divergence operator with respect to μ, i.e., $\delta = \nabla^\star$, and the adjoint is calculated with respect to μ. By the symmetry of μ, it is easy to see that $\sigma(0) = 0$, and since $P_t g$ is again log-concave ([2]), $(\sigma(x), x) \leq 0$, for any $x \in \mathbb{R}^n$, where $(\cdot,\cdot)$ denotes the scalar product in $\mathbb{R}^n$. Define now $\sigma_t(x)$ as

$$\sigma_t(x) = \frac{\sigma(x)}{t + (1-t)g(x)}\,. \tag{3}$$

From what we have explained above, σ_t is a dissipative vector field, consequently the differential equation defined by

$$\phi_{s,t}(x) = x - \int_s^t \sigma_\tau(\phi_{s,\tau}(x)) d\tau \tag{4}$$

defines a diffeomorphism of $W = \mathbb{R}^n$ with the property $|\phi_t(x)| \geq |x|$. Let

$$\Lambda_{s,t} = \exp \int_s^t (\delta\sigma_\tau)(\phi_{s,\tau}) d\tau.$$

We then have

$$\int F \circ \phi_{s,t} \Lambda_{s,t} d\mu = \int F d\mu$$

for any $F \in C_b(W)$ ([5]). Let $H_s(t,x)$ be the function defined as

$$H_s(t,x) = \Lambda_{s,t}(x)\left[t + (1-t) g \circ \phi_{s,t}(x)\right].$$

It is easy to see that

$$\frac{d}{dt} H_s(t,x) = 0\,.$$

Consequently we have

$$\Lambda_{s,1} = [s + (1-s)g(x)].$$

Then we have, from the Jacobi formula:

$$\int 1_A \circ \phi_{s,1}(x)(s + (1-s)g(x)) d\mu(x) = \mu(A)\,.$$

On the other hand,

$$\int 1_A \circ \phi_{s,1}(x)(s + (1-s)g(x) d\mu(x) \leq \int 1_A(x)(s + (1-s)g(x) d\mu(x),$$

where the inequality originates from the dissipativity of σ_t which implies $|\phi_1(x)| \geq |x|$ and from the hypothesis about A: $1_A \circ \phi_1^{-1} \geq 1_A$. The proof follows since $s \in (0,1)$ is arbitrary and since $g = \frac{1}{\mu(B)} P_\varepsilon 1_B$ converges to $\frac{1}{\mu(B)} 1_B$ as $\varepsilon \to 0$. □

References

[1] B. Dacorogna and J. Moser: On a partial differential equation involving the Jacobian determinant. *Ann. Inst. Henri Poincaré, Analyse non linéaire*, **7** (1990), 1–26.

[2] D. Feyel and A.S. Üstünel: The notion of convexity and concavity on Wiener space. *Journal of Functional Analysis*, **176** (2000), 400–428.

[3] L. D. Pitt: A Gaussian correlation inequality for symmetric convex sets. *The Annals of Probability*, **5** (1977), 470–474.

[4] A.S. Üstünel: *Introduction to Analysis on Wiener Space*. Lecture Notes in Math. Vol. 1610. Springer, 1995.

[5] A.S. Üstünel and M. Zakai: *Transformation of measure on Wiener space*. Springer Verlag, 1999.

Départment Résaux
46 rue Barrault
75634 Paris Cedex 13
France

GPSR Compliance

The European Union's (EU) General Product Safety Regulation (GPSR) is a set of rules that requires consumer products to be safe and our obligations to ensure this.

If you have any concerns about our products, you can contact us on ProductSafety@springernature.com

In case Publisher is established outside the EU, the EU authorized representative is:

Springer Nature Customer Service Center GmbH
Europaplatz 3
69115 Heidelberg, Germany

Batch number: 10577884

Printed by Printforce, the Netherlands